APPLIED BIOPOLYMER TECHNOLOGY AND BIOPLASTICS

Sustainable Development by
Green Engineering Materials

APPLIED BIOPOLYMER TECHNOLOGY AND BIOPLASTICS

Sustainable Development by
Green Engineering Materials

Edited by

Neha Kanwar Rawat, PhD
Tatiana G. Volova, DSc
A. K. Haghi, PhD

First edition published 2021

Apple Academic Press Inc.
1265 Goldenrod Circle, NE,
Palm Bay, FL 32905 USA
4164 Lakeshore Road, Burlington,
ON, L7L 1A4 Canada

CRC Press
6000 Broken Sound Parkway NW,
Suite 300, Boca Raton, FL 33487-2742 USA
2 Park Square, Milton Park,
Abingdon, Oxon, OX14 4RN UK

First issued in paperback 2021

Library and Archives Canada Cataloguing in Publication

Title: Applied biopolymer technology and bioplastics : sustainable development by green engineering materials / edited by Neha Kanwar Rawat, PhD, Tatiana G. Volova, DSc, A.K. Haghi, PhD.
Names: Rawat, Neha Kanwar, editor. | Volova, Tatiana G., editor. | Haghi, A. K., editor.
Description: Includes bibliographical references and index.
Identifiers: Canadiana (print) 20200363255 | Canadiana (ebook) 20200363328 | ISBN 9781771889216 (hardcover) | ISBN 9781003045458 (ebook)
Subjects: LCSH: Biopolymers. | LCSH: Biodegradable plastics. | LCSH: Green technology. | LCSH: Sustainable engineering.
Classification: LCC TP248.65.P62 A67 2021 | DDC 572/.33—dc23

Library of Congress Cataloging-in-Publication Data

....

CIP data on file with US Library of Congress

....

ISBN: 978-1-77188-921-6 (hbk)
ISBN: 978-1-77463-774-6 (pbk)
ISBN: 978-1-00304-545-8 (ebk)

About the Editors

Neha Kanwar Rawat, PhD
Researcher, Materials Science Division,
CSIR – National Aerospace Laboratories, Bangalore, India

Neha Kanwar Rawat, PhD, is a recipient of a prestigious DST Young Scientist Postdoctoral Fellowship and is presently a researcher in the Materials Science Division, CSIR – National Aerospace Laboratories, Bangalore, India. She received her PhD in chemistry from Jamia Millia Islamia (A Central University), India. Her main interests include nanotechnology-nanostructured materials synthesis and characterization, her main focus comprising green chemistry, novel sustainable chemical processing of nano-conducting polymers/nanocomposites, conducting films, ceramics, silicones, and matrices: epoxies, alkyds, polyurethanes, etc. She also pursues her interest in using new technology in areas that include electrochemistry, organic-inorganic hybrid nanocomposites, and protective surface coatings for corrosion inhibition and MW shielding materials. She has published numerous peer-reviewed research articles in journals of high repute. Her contributions have led to many chapters in international books published with the Royal Society of Chemistry, Wiley, Elsevier, Apple Academic Press, Nova U.S., and many others in progress. She has been working on many prestigious research and academic fellowships in her career. She is a member of many groups, including the Royal Society of Chemistry and the American Chemical Society (USA) and a life member of the Asian Polymer Association.

Tatiana G. Volova, DSc
Professor and Head, Department of Biotechnology,
Siberian Federal University, Krasnoyarsk, Russia

Tatiana G. Volova, DSc, is a Professor and Head in the Department of Biotechnology at Siberian Federal University, Krasnoyarsk, Russia. She is the creator and head of the Laboratory of Chemoautotrophic Biosynthesis at the Institute of Biophysics, Siberian Branch of the Russian Academy of

Sciences. Professor Volova is conducting research in the field of physico-chemical biology and biotechnology and is a well-known expert in the field of microbial physiology and biotechnology. Dr. Volova has created and developed a new and original branch in chemoautotrophic biosynthesis, in which the two main directions of the 21st century technologies are conjugate, hydrogen energy, and biotechnology. Under the guidance of Professor Volova, the pilot production facility of single-cell protein, utilizing hydrogen had been created and put into operation. The possibility of involvement of man-made sources of hydrogen into biotechnological processes as a substrate, including synthesis gas from brown coals and vegetable wastes, was demonstrated in her research. She had initiated and deployed in Russia comprehensive research on microbial degradable bioplastics; the results of this research cover various aspects of biosynthesis, metabolism, physiological role, structure, and properties of these biopolymers and polyhydroxyalkanoates (PHAs), and have made a scientific basis for their biomedical applications and allowed them to be used for biomedical research. Professor Volova is the author of more than 300 scientific works, including monographs, inventions, and a series of textbooks for universities.

A. K. Haghi, PhD

Professor Emeritus of Engineering Sciences, Former Editor-in-Chief, International Journal of Chemoinformatics and Chemical Engineering and Polymers Research Journal; Member, Canadian Research and Development Center of Sciences and Culture

A. K. Haghi, PhD, is the author and editor of 200 books, as well as 1000 published papers in various journals and conference proceedings. Dr. Haghi has received several grants, consulted for a number of major corporations, and is a frequent speaker to national and international audiences. Since 1983, he served as a professor at several universities. He is former Editor-in-Chief of the *International Journal of Chemoinformatics and Chemical Engineering* and *Polymers Research Journal* and is on the editorial boards of many international journals. He is also a member of the Canadian Research and Development Center of Sciences and Cultures (CRDCSC), Montreal, Quebec, Canada.

Contents

Contributors

Raluca Marinica Albu
"Petru Poni" Institute of Macromolecular Chemistry, Laboratory of Physical Chemistry of Polymers, 41A Grigore Ghica Voda Alley, Iasi – 700487, Romania

Jimsher Aneli
Institute of Machine Mechanics, 10, Mindeli Str., Tbilisi – 0186, Georgia, E-mail: jimaneli@yahoo.com

P. Muhamed Ashraf
ICAR Central Institute of Fisheries Technology, Cochin – 682029, Kerala, India, E-mail: ashrafp2008@gmail.com

Akshada A. Bakliwal
Department of Pharmaceutics, Sandip Institute of Pharmaceutical Sciences, Nashik, Maharashtra, India

Andreea Irina Barzic
"Petru Poni" Institute of Macromolecular Chemistry, Laboratory of Physical Chemistry of Polymers, 41A Grigore Ghica Voda Alley, Iasi – 700487, Romania

Brahmananda Chakraborty
High Pressure and Synchrotron Radiation Physics Division,

Yogesh Chendake
Department of Chemical Engineering, Bharati Vidyapeeth (Deemed to be) University, College of Engineering, Pune, Maharashtra, India, E-mail: yjchendake@bvucoep.edu.in

Sanjit Das
Material Science Section, Hari Shankar Singhania Elastomer and Tyre Research Institute (HASETRI), Plot No 437, Hebbal Industrial Area, Mysore, Karnataka – 570016, India, E-mail: sanjit.das@jkmail.com

Aleksey V. Demidenko
Siberian Federal University, 79 Svobodnyi Av., Krasnoyarsk – 660041, Russia; Institute of Biophysics SB RAS, Federal Research Center "Krasnoyarsk Science Center SB RAS," 50/50 Akademgorodok, Krasnoyarsk – 660036, Russia

Supriya Dhume
Department of Chemical Engineering, Bharati Vidyapeeth (Deemed to be) University, College of Engineering, Pune, Maharashtra, India

Saikat Das Gupta
Material Science Section, Hari Shankar Singhania Elastomer and Tyre Research Institute (HASETRI), Plot No 437, Hebbal Industrial Area, Mysore, Karnataka – 570016, India, E-mail: saikat.dasgupta@jkmail.com

Athira John
Center for Biopolymer Science and Technology (CBPST), CIPET, Kochi, Kerala, India

Evgeniy G. Kiselev
Siberian Federal University, 79 Svobodnyi Av., Krasnoyarsk – 660041, Russia; Institute of Biophysics SB RAS, Federal Research Center "Krasnoyarsk Science Center SB RAS," 50/50 Akademgorodok, Krasnoyarsk – 660036, Russia

Cristian Logigan
Chemical Company SA, 14 Chemistry Bdv., Iasi – 700293, Romania

Grigor Mamniashvili
Andronikashvili Institute of Physics of Ivane Javakhishvili Tbilisi State University,
6, Tamarashvili Str., Tbilisi – 0177, Georgia

Swapnali A. Patil
Department of Pharmaceutics, Sandip Institute of Pharmaceutical Sciences, Nashik, Maharashtra, India

Ananthu Prasad
International and Inter-University Center for Nanoscience and Nanotechnology (IIUCNN),
Mahatma Gandhi University, Kottayam, Kerala, India

Neha Kanwar Rawat
CSIR – National Aerospace Laboratories Bangalore, Karnataka, India

S. Roopa
Department of Polymer Science and Technology, Sri Jayachamarajendra College of Engineering,
JSS Science and Technology University, Mysore – 570006, India, E-mail: roopasm2000@jssstuniv.in

Gopal Sanyal
Mechanical Metallurgy Division, Bhabha Atomic Research Center, Trombay, Mumbai – 400085, India

Hirak Satpathi
Material Science Section, Hari Shankar Singhania Elastomer and Tyre Research Institute (HASETRI),
Plot No 437, Hebbal Industrial Area, Mysore, Karnataka – 570016, India, E-mail: hirak@jkmail.com

Ekaterina I. Shishatskaya
Siberian Federal University, 79 Svobodnyi Av., Krasnoyarsk – 660041, Russia;
Institute of Biophysics SB RAS, Federal Research Center "Krasnoyarsk Science Center SB RAS,"
50/50 Akademgorodok, Krasnoyarsk – 660036, Russia

Swati G. Talele
Department of Pharmaceutics, Sandip Institute of Pharmaceutical Sciences, Nashik, Maharashtra, India,
E-mail: swatitalele77@gmail.com

Sabu Thomas
Siberian Federal University, 79 Svobodnyi Av., Krasnoyarsk – 660041, Russia;
International and Interuniversity Center for Nano Science and Nano Technology,
Mahatma Gandhi University, Kottayam, Kerala, India

Tatiana Gr. Volova
Siberian Federal University, 79 Svobodnyi Av., Krasnoyarsk – 660041, Russia;
Institute of Biophysics SB RAS, Federal Research Center "Krasnoyarsk Science Center SB RAS,"
50/50 Akademgorodok, Krasnoyarsk – 660036, Russia, E-mail: volova45@mail.ru

Abbreviations

ACL	anterior cruciate ligaments
AgNPs	silver nanoparticles
ASW	artificial/natural seawater
ATP	adenosine triphosphate
BLYP	Becke-Lee-Yang-Parr
BMIAc	1-butyl-3-methylimidazolium acetate
BmimCl	1-butyl-3-methylimidazolium chloride
BOD	biochemical oxygen demand
CC	climate change
CdO	cadmium oxide
CHT/SS	chitosan/silk sericin
CMC	critical micelle concentration
CMPs	conjugated microporous polymers
CMT	critical micelle temperature
CNCs	cellulose nanocrystals
CNF	cellulose nanofibers
CNHs	carbon nanohorns
CNTs	carbon nanotubes
CO_2	carbon dioxide
COFs	covalent organic frameworks
CS-g-PNIPAAm	chitosan-graft-poly(N-isopropyl acrylamide)
CTFs	covalent triazine frameworks
CVD	chemical vapor deposition
DADS	diallyl disulfide
DCMH	doxycycline monohydrate
DFT	density functional theory
DMF	N,N-dimethylformamide
DMSO	organic solvents-dimethyl sulfoxide
DoE	department of energy
DSC	differential scanning calorimetry
DVA	dynamically vulcanized alloy
DWCNTs	double-walled carbon nanotubes
ECM	extracellular matrix
EGF	epidermal GF

EmimAc	1-ethyl-3-methylimidazolium acetate
ENR	epoxidized natural rubber
E-SBR	emulsion styrene-butadiene rubber
EtOH	ethanol
ETRMA	European Tire and Rubber Manufacturer's Association
FFA	free fatty acids
FO	forward diffusion
FTIR	Fourier transform infrared
GGA	generalized gradient approximation
GT	gelatin
H_2	hydrogen
HCPs	hyper cross-linked polymers
HDS	highly dispersive silica
HFC	hydrogen fuel cells
HPAM	hydrolyzed polyacrylamide
HPC	hydroxypropyl cellulose
HPMS	hydroxypropylmethylcellulose
HSM	high strength Mixer
IBET	integrated biology, English, and technology
ICE	internal combustion engine
IgG	immunoglobulin
IUPAC	International Union of Pure and Applied Chemistry
IUU	illegal, unreported, and unregulated
LA	lactic acid
LDA	local density approximation
MD	membrane distillation
MeCN	acetonitrile
MF	microfiltration
MFCs	microbial fuel cells
MgO	magnesium oxide
MOFs	metal-organic frameworks
MWCNTs	multi-walled nanotubes
NF	nanofiltration
NR	natural rubber
PAFs	porous aromatic frameworks
PAH	polyaromatic hydrocarbons
PBE	Perdew-Burke-Ernzerhof
PbO	lead oxide
PCA	polycyclic aromatics

PCL	polycaprolactone
PDMS	polydimethylsiloxane
PECs	polyelectrolyte complexes
PHAs	polyhydroxyalkanoates
PI	process intensification
PIMs	polymers of intrinsic microporosity
PNC	polymer nanocomposites
POP	porous organic polymer
PP	polypropylene
PPO	poly(propylene oxide)
PS	polystyrene
PTFE	polytetrafluoroethylene
PV	pervaporation
PVA	polyvinyl alcohol
PVB	polyvinyl butyral
PVI	prevulcanization inhibitor
RCIM	Russian Collection of Industrial Microorganisms
RH	hydrodynamic radius
RO	reverse osmosis
RRI	Rubber Research Institute
SAXS	small-angle x-ray scattering
SEM	scanning electron microscope
SF	silk fibroin
S-SBR	solution styrene-butadiene rubber
STE	science-technology-engineering
SWCNT	single-walled carbon nanotube
TBR	truck bus radial
TBT	tributyltin
TBzTD	tetrabezylthiuram disulfide
TE	tissue engineering
TEO	thyme essential oil
TGA	thermogravimetric analysis
UF	ultrafiltration
USTMA	U.S. Tire Manufacturing Association
VP	vapor permeation
WPC	whey protein concentrates
WPU	waterborne polyurethane
XRD	x-ray diffraction
ZnO	zinc oxide

Preface

With growing concern for the environment and the rising price of crude oil, there is increasing demand for non-petroleum-based polymers from renewable resources. Recognizing emerging developments in biopolymer systems research with fully updated and expanded chapters, this book brings together a number of key biopolymer and bioplastic topics in one place for a broad audience of engineers and scientists, especially those designing with biopolymers and biodegradable plastics, or evaluating the options for switching from traditional plastics to biopolymers.

Bioplastics are, on the one hand, bio-based plastics (produced from renewable resources) and, on the other hand, bioplastics may well be biodegradable plastics. Bio-based plastics have a huge market and a wide range of applications and then have become an increased in economy and research. In today's world, bioplastics are becoming increasingly prominent owing mainly to scarcity of oil, an increase in the cost of petroleum-based commodities, and growing environmental concerns with the dumping of non-biodegradable plastics in landfills.

Bioplastics are more sustainable because they can break down in the environment faster than fossil-fuel plastics, which can take more than 100 years. Some, but not all, bioplastics are designed to biodegrade. Bioplastics that are designed to biodegrade can break down in either anaerobic or aerobic environments, depending on how they are manufactured.

The volume provides insight into the diversity of polymers obtained directly from, or derived from, renewable resources. The book highlights the importance and impact of eco-friendly green biopolymers and bioplastics, both environmentally and economically. The contents of this book will prove useful for students, researchers, and professionals working in the field of green technology.

Green chemistry concepts are becoming more and more important in the production of polymer materials as a result of increasing demand for green and renewable products on various markets. Chapter 1 is focused on presentation the importance of rheology as a tool for predicting the optimum conditions of processing of the materials containing green polymer as main or secondary components. Polymer solutions in less toxic solvents have specific flow behavior as a function of the composition of the analyzed system. The viscoelastic character is also a key factor in shaping green polymers in the form of the desired product. The practical importance of some macromolecular systems containing green polymers is also briefly reviewed. Current developments in this field are described from a rheological point of view.

In Chapter 2, the self-assembling processes were studied in the magnetic polymer nanocomposites (PNC) synthesized based on the carbon nanoparticles doped with cobalt nanoclusters, synthesized by a unique CVD technology developed by the authors, under the combined effect of magnetic field and heating. These processes took place with the diffusion of magnetic nanoparticles stimulated by the combined effect of the outer steady magnetic field and heating. The obtained magnetic polymer nanocomposites have good electrical and adhesive properties and are promising for potential practical applications in magneto-electronics.

In Chapter 3, it is shown that microbial fuel cell (MFC) is one of the recent advanced sciences of using microbes for the generation of electricity. Microbial fuel cell undergoes an aerobiotic ferment of organic and inorganic matter to digest fuel which is obtained in energy form. This methodology is not so applicable due to expensiveness. MFC is an electrochemical science where chemical energy is converted to electrical energy using different microbes. In other words, MFC can also be titled as cellular respiration. The basic concept in is the reduction-oxidation respiration in the microbial culture which generates energy in electrical form. The movement of electrons produces a potential difference which in turn helps to generate fuel. A path of hope for MFC conceptual is progressively studied to develop a new electrode in recent days. MFC is into a study to upgrade the concept with cost-effectiveness and helpful application of the technique in a new glance.

Tire industry is basically a slow-changing industry, however, in the present era of fast change, a number of environmental and sustainability-related issues are impacting the tire industry to a large extent. Non-pneumatic tire, tires for electric vehicles, tires with sensors is talk of the day to meet the future requirements of the automobile industries. All these tires need to have fuel efficiency, safety at high speed, and low noise. Tire designers and material scientists are jointly involved in tire design, material design to achieve targeted tire performance properties. Material scientists are focusing their research mainly on sustainable materials as an alternative to present fossil-based material. In Chapter 4 authors shared different research works that are being carried out on sustainable materials particularly for the tire industry. Reuse-regenerate-recycle is one of the research areas to reduce the consumption of virgin fossil fuel-based materials. Natural rubber (NR) is one of the commonly used rubbers in the tire industry. It is a naturally occurring material and helps the tire industry to achieve sustainability in the long run. However, the imbalance between the demand and supply of Natural rubber, worldwide scientists are doing research on Guayule, Dandelion to obtain a suitable alternative of natural rubber. With the introduction of stringent requirements of tire labeling criteria by the European Union, materials scientists are bound to use the alternative of carbon black reinforcing filler. One of the major alternatives of carbon black is silica which is also an eco-friendly sustainable material. Synthetic rubbers using plant-derived materials or biomass (agricultural waste) are came into the picture, Liquid Farnesene rubber is also another newly introduced polymer from bio origin. Research efforts in developing microbes that can grow an isoprene monomer used to make a mimic NR is also undergoing. Fillers which are the next bulk consuming material after polymer are being explored from non-petroleum-based origin like silica, clay, and from bio-origin like rice husk ash, cornflower starch, bagasse; eggshell derived nano-calcium carbonate. Coupling agents like Baker's yeast as an alternate of TESPT. The process aids from non-petroleum-based origin like soybean oil, orange oil, canola oil, sunflower oil, neem oil, etc., are also under the different stages of research for application in tire. Apart from that, many resins from a natural source are used to replace petroleum-based process oil and also to enhance tire performance. Cellulose-based synthetic fibers are of profound interest to replace synthetic fiber in tire reinforcement. Researchers are working to get anti-oxidant from flowers, fruits peels, and food wastes to replace petro-based antioxidants. Tea polyphenols as anti-oxidant have already found potential for the rubber compound. Investigations are also being done for a single compounding ingredient that would possess several functions and thus the term surfactant-accelerator-processing aid (SAPA) or termed as multi-functional additives (MFA).

The main scope of Chapter 5 is to show that the biopolymer silk as a recognizable implement in the biomedical field is manifested in nanotechnology due to its promising avenues. Silk has been esteemed and caught the attention of the human race, by virtue of its lustrous appeal since long ago. Therefore, it rules the textile market across the globe even today. But apart from its aesthetic value researchers have attributed some medicinal values to silk in the recent past and its scope is expected to shoot up shortly. It was surprising that, even sericin which was once considered the waste material in the silk industry had got special consideration due to its unique properties. Silk proteins were found to have excellent biocompatibility and slow biodegradability which provides it an unavoidable role in biomedical applications. The structure composition and current researches like wound healing, tissue engineering (TE), and the emerging biomedical applications of silk in its various forms like films, nanofiber, hydrogels, and particles were discussed. Silk by undergoing major and minor chemical modification or by combining with different synthetic or natural polymers can work wonders in the biomedical field. There is still a vast world of silk unexplored; the researchers have to research these hidden capable avenues and their applications.

Biofouling in aquaculture cages has been an important issue for the farmers since its management incurs a huge sum of money and labor. Antifouling strategies failed due to the diversity of organisms are different from region to region and increased concern about pollution in the water bodies. Chapter 6 describes the strategies to combat biofouling using nanomaterials coated over a polyaniline coated polyethylene and its efficiency towards antifouling. Green synthesis of nanomaterial incorporated hydrogel and mixed charged polymeric hydrogel over polyethylene aquaculture cage nets against biofouling also described. Brief description of the environmental impact assessment of nanomaterials treated nets in the marine environments included at the end.

In Chapter 7, mesoporous SiO_2, gender, climate change (CC), and science in parliament investigated.

In Chapter 8, two approaches have been investigated in order to reduce the cost and increase the availability of the use of biodegradable polymers of microbiological origin (polyhydroxyalkanoates, PHA): (1) expanding the raw material base and attracting glycerol as a carbon substrate (large-tonnage waste from the production of biodiesel) and (2) filling the polymer with natural materials for industrial use. An effective technology for the synthesis of destructible polymers was developed using a new and more affordable substrate compared to sugars-glycerol. The *Cupriavidus eutrophus* B-10646 strain was studied as a product of polymers on glycerin, and the kinetic and production characteristics of the strain providing polymer yields on glycerol of various cheat, comparable with the process on sugars, were studied. Replacing sugars with glycerol provides a reduction in the cost of a carbon substrate in the production of PHA. To increase the availability of PHA and their technical applications, mixtures of PHA with natural materials were formed, the structure and properties, as well as the patterns of degradation in the soil, were studied. The possibility of long-term functioning of mixed forms in the soil with a gradual release of active substances (herbicides and fungicides) and the possibility of providing plants with protective equipment during the growing season are shown. Both approaches are effective for enhancing accessibility and expanding the scope of application of perspective green plastics-destructible PHA.

Green technology is that to operate the ecological skills for the development and application of merchandise, instrumentality, and systems to conserve the natural resources and surroundings. This field is moderately innovative. It has developed rapidly as people have become aware of the detrimental effects of environmental variation including global warming and adverse effects on the living beings. It is believed that growth in green technology will lead to worldwide, sustainable, and economic powers that impact economics, societies, and way of life in the future. The application of green technology has multiple challenges while it can provide many opportunities during real-life chemical processes. The membrane technology can provide an economical pathway to many of these challenges. Here applicability of membranes during green technology is discussed. Over the past few years, the membrane separation process has become one of the major separation technologies. It is rapidly becoming an important part of green technology due to some of its benefits. In membrane processes do not require the addition of chemicals, it works on physical separation. This improves the feasibility of recycling the recovered chemicals or using them in further processes. Additionally, membrane technology has relatively low energy requirements, easy to design and integrated nature may advance, and can be integrated with current processes. This gives benefits of reduced production cost, equipment size, energy consumption, and waste generation and improving controls and process flexibility.

Chapter 9 would consider two aspects of membrane technologies for process enhancements; as process integration and effluent treatment. Process integration would be focused on for improvements in economical separation, recovery, and recycle; this would reduce the generation of waste and effluent component generation, thus enhancing the process financial system. Membrane separation technologies are very helpful in effluent treatment. It is investigated for a reduction in effluent volume and its treatment for components to recover. The work starts by presenting an overview of membrane operations. It would be followed for the application of membrane processes for process improvement and *in situ* recovery of components. This would be followed by the effluent treatment aspects. Such recovery of components harmful to the environment and their further utilization by membrane processes would increase the process economy and product purity. To get superior removal of contaminants compared to conventional technologies, the membrane technology would provide better support to the system. The conventional technologies frequently required continuous modification of conditions and the use of chemicals. This means a high level of operator involvement and the risk of contaminants in the effluent.

One of the greatest challenges of this century is to find an alternative fuel for vehicular propulsion which can replace widely used fossil fuels. Hydrogen is one of the finest choices due to its clean energy and highly plentiful in nature. And also, it has the highest energy value per unit weight ($142 \ MJ \ kg^{-1}$) compared to liquid hydrocarbons ($47 \ MJ \ kg^{-1}$). The conventional hydrogen storage technologies, namely, high-pressure tank and liquid state storage, are not applicable due to large size and higher energy cost for liquefaction. Solid-state storage may become a viable technology provided the storage medium can absorb a large amount (~6.5 wt%) of hydrogen and can release them easily as recommended by the department of energy (DoE), USA. Recently, carbon nanomaterials have drawn immense attention from researchers due to their high surface area, high electrical and thermal conductivity, and high mechanical strength. Although for pure carbon nano-structures, the physisorption of hydrogen is negligible at room temperature, the functionalized carbon nanomaterials are the promising candidates for hydrogen storage media at ambient conditions. This chapter describes the issues and challenges for hydrogen storage in functionalized carbon nano-materials. It will highlight the bonding, charge transfer mechanism, and hydrogen storage capability of novel carbon nanostructure (carbon nanotube, graphene, and graphyne).

Polymers are also potential hydrogen storage materials due to their pure organic nature, tunable structures, and large scale synthesis. Chapter 10 discusses about different types of polymers such as conjugated micro porous polymers, porous polymers networks, polymers of intrinsic microporosity (PIM), hyper cross-linked polymers, and their hydrogen storage capabilities.

Materials developed on the basis of green polymers are of major impor-tance not only for the keeping a clean environment, but also for creating of advanced products with architecture controlled at the micro- and nano-scale level. The last chapter reviews the progress made in the field of self-assem-blies in green macromolecular systems. The first short section presents some theoretical aspects and the most used synthesis and preparation routes of macromolecular assemblies of biopolymers. Then, the commonly employed characterization techniques of such soft nanomaterials. The newest developments in green polymer self-assemblies are described emphasizing their practical importance. The presented research examples prove the range of possibilities and future trends in self-assembly processes that will significantly contribute to the creation of materials nanoarchitectonics. The green materials, here the topic under analysis, cover several application fields.

CHAPTER 1

Processing Relation to Rheological Behavior of Some Systems Containing Green Polymers

ANDREEA IRINA BARZIC

"Petru Poni" Institute of Macromolecular Chemistry,
Laboratory of Physical Chemistry of Polymers,
41A Grigore Ghica Voda Alley, Iasi – 700487, Romania

ABSTRACT

Green chemistry concepts are becoming more and more important in the production of polymer materials as a result of increasing demand for green and renewable products in various markets. The chapter is focused on the presentation of the importance of rheology as a tool for predicting the optimum conditions of processing of the materials containing green polymer as main or secondary components. Polymer solutions in less toxic solvents have specific flow behavior as a function of the composition of the analyzed system. The viscoelastic character is also a key factor in shaping green polymers in the form of the desired product. The practical importance of some macromolecular systems containing green polymers is also briefly reviewed. Current developments in this field are described from a rheological point of view.

1.1 INTRODUCTION

The serious consequences of our planet the pollution have drawn the attention of the scientific community and industrial market towards green chemistry and related products [1–4]. In the 1990s, the International Union of Pure and

Applied Chemistry (IUPAC) introduced green chemistry as a field that is connected to the "design of chemical products and processes that reduce or eliminate the use or generation of substances hazardous to humans, animals, plants, and the environment" [5]. In other words, green chemistry attempts to diminish and prevent pollution at its source. Based on these concepts, the research tendencies in material science begin to change opening new perspectives concerning synthesis techniques and processing procedures [6–8]. In some cases, the material's level of performance is not as high as expected, but opens new paths for a healthier environment.

When it comes to polymers, there is a slight difference between *natural* and *green* ones [5]. As stated by their name, *natural polymers* (or biopolymers) are represented by macromolecules that can be encountered naturally or they are synthesized by living organisms. Among them, one can mention cellulose, silk, chitin, DNA, and protein. Another definition of natural polymers refers to macromolecular compounds which can be obtained from raw materials that came across in nature [5]. Each year, around 300 million tons of plastics are manufactured, among which natural polymers represent less than 1%, but their production tends to increase [5]. For instance, the food packaging, cosmetic, medical, pharmaceutical, and beverage industries require large amounts of biopolymers, supporting the development of the natural polymers market. The latter is dominated by cellulose ethers, starch-based materials, fermentation polymers, exudates, protein-derived polymers, and marine polymers [5].

On the other hand, when referring to *green polymers* one should have in mind macromolecular compounds achieved through green (or sustainable) chemistry procedures [8–10]. In order to reduce pollution during polymer preparation, one should use adapted techniques that minimize the contamination of the environment. For this reason, natural polymers are usually considered green polymers [5].

The 21st century brought again in attention renewable polymers and major thrust towards the design of various bio-materials of macromolecular nature. For a close control of the pursued properties, a careful attention of each component of the final material is performed. So, there is a gradual transition from petrochemistry to bioeconomy [11], making inevitable the development of the green polymer market. Analyzing these aspects economically, it is obvious that remaining oil resources will be achieved with higher expenses, whereas the growing worldwide energy demands cannot be fully satisfied. Such aspects might have undesired effects on the cost-effectiveness and competitiveness of plastics [5]. Switching chemical raw material manufacturing to renewable resources might secure plastics production against a potential future oil crisis.

Another important issue that could contribute in favor of green plastics are the environmental problems that arise from plastics wastes from numerous industries, like electronics, packaging, paints, and so on [12–14]. Thus, the consumers concerns related to global warming can be solved by replacing the current products with sustainable and 'green' ones. Moreover, several states formulated environmental legislation and regulations which support the thrive of eco-friendly products with a reduced carbon footprint.

In the synthesis and processing of macromolecular compounds, the green principles are referring to [5, 7, 15]:

- A large amount of raw material in the product;
- A clean (lack of waste) production process;
- No use of supplementary substances like organic solvents;
- High energy efficiency in manufacturing;
- Utilization of renewable resources and renewable energy;
- No generation of health and environmental hazards;
- Raised safety standards;
- Reduced carbon footprint;
- Controlled product life cycles with effective waste recycling.

The utilization of renewable resources for green polymer preparation should not compete with food production, but more important must not encourage farming or deforestation, while most not involve transgenic plants or genetically adapted bacteria [5]. The latter aspects are crucial for human health since biodegradable polymers should not generate inhalable spores or nanoparticles.

Having all these in view, one may distinguish three main routes to produce renewable plastics [5, 7, 15]:

- Utilization of biomass and/or carbon dioxide to prepare 'renewable oil' and green reactants for highly resource- and energy-effective polymer synthesis procedures.
- Use of living cells, which are viewed as solar-powered chemical reactors through genetic engineering and biotechnology techniques to generate bio-based polymers.
- Via activation and polymerization of carbon dioxide.

To ensure and consolidate the progress in green polymers, researchers, and manufacturers work together and make known their ideas at specific scientific events dedicated to production of green/renewable polymers and bioplastics and from various resources, such as plants and bio-refineries, algae, waste, and carbon dioxide [5, 15]. Nowadays this is viewed in the emerging biofuels, like bioethanol, which is made by fermentation of sugar achieved from cellulose. This biofuel is also utilizable in the synthesis of green polyethylene, which is entirely recyclable. The energy-efficient processing techniques allow the transformation of the biomass into renewable coal and oil. There are certain agricultural and forestry residues already which can be used to prepare renewable monomers.

Carbohydrates, proteins, and polyesters are chemically adapted for green material processing and applications. Natural fibers are considered outstanding reinforcement agents for thermosets and thermoplastics. As an example, micro-fibrillated cellulose has been used in nanocomposites with polymer matrix in many applications, including in medical implants. Lignin is considered to be a renewable energy source in paper production, and also as filler for several polymers and rubbers. Thermoplastic lignin combined with natural fibers (e.g., Arboform) preserves the advantages of wood and synthetic thermoplastic materials [5]. Bio-hybrids have been prepared by the introduction of starch in polyolefins or in compostable polyesters. Biopolymers, such as chitosan and polylactic acid display numerous medical applications. Phosphoproteins, like Casein, have been proved to be good binders or adhesives. Renewable monomers are already substituting for "oil-made" monomers. This is widely present in plastic bottles that currently are made of bio-based materials which can be recycled. Generally speaking, it is not so easy for polymer production to fulfill the demands of green chemistry, but not impossible. Thus, green polymer preparation routes must involve non-toxic solvents or even solvent-free procedures, no byproduct formation and efficient use of renewable resources [5, 15].

The processing of the materials containing green polymer as main or secondary components in the shape of the desired products especially starts from solution or melt state [16–18]. A large category of polymers degrade prior to melting so solution processing remains the only path toward finite product manufacturing. In both situations, it is of paramount importance to know the flow behavior and viscoelasticity of the system before processing. One of the most sensitive tools for analysis of polymer solutions (or melts) in the semi-dilute concentration domain is *rheology*. This technique allows extraction of a large amount of information that facilitate better understanding of the response of the polymer sample to shear deformation, which is often encountered during the processing stage [19].

This chapter is devoted to present the state-of-art of rheological characterization of solution containing green polymers and the applicability of such systems in various industries. Also, a short presentation concerning the environmental impact of these materials is reviewed.

1.2   SOLUTION RHEOLOGY CORRELATION WITH POLYMER PROCESSING

Polymer dissolution is quite distinct than that of low molecular weight materials since it involves two stages: solvent diffusion and chain disentanglement [20]. When an amorphous and glassy polymer (with no cross-links) is placed in a thermodynamically compatible solvent, the solvent begins to diffuse among the macromolecules. Owing to the plasticization of the polymer in the presence of the solvent, a gel-like swollen layer is created along with two separate interfaces: one between the glassy polymer and gel layer and the other between the solvent molecules and the polymer gel. After a period of time, there are three possible situations:

- The first case refers to fully polymer dissolution;
- The second probability is linked to a partial dissolution of the polymer; and
- In the last case, the polymer cracks and no gel layer is produced.

When macromolecules are dissolved, they tend to adopt a variety of conformations [21–23]. A large number of chain shape is enabled by many internal rotations which can form through simple carbon-carbon bonds creating several rotational isomers. As a result of energy restrictions, not all conformations display a similar probability of occurrence [22]. Therefore, the most stable conformations are the ones that prevail in solution. If the macromolecules are highly flexible they often change their conformation and instead of having a linear form in solution, the chains adopt a very typical conformation, named random coil. It is worth mentioning that the random coil would not remain the same after the polymer has been dissolved. This is because the solution behavior of polymers is strongly affected by several factors, such as [21–23]:

- Polymer structure, composition, and conformation;

- Molecular weight and polydispersity index;
- Thermodynamic quality of the solvent;
- Polymer-solvent interactions;
- Temperature.

The dissolution process can be influenced by the chain chemistry, composition, and stereochemistry [20]. The polymer dissolves either by displaying a thick swollen layer or by experiencing cracking, as a result of relieving ability of the osmotic pressure stress those results in the polymer. Hence, the nature of the polymers and dissimilarities in free volume and chain segment stiffness are ascribed to the properties variations from polymer to polymer.

It is widely known that the dissolution rate is lower as the polymer molecular weight is higher, while the polydisperse polymers dissolved about twice as fast as monodisperse ones of identical molecular weight [20, 24–26].

In a binary solution prepared from a polymer-solvent pair having similar solubility parameters, the forces acting between chain segments become smaller than those occurring between the polymer and the solvent [24]. In such a system, the random coil adopts an unfolded conformation. In a less compatible solvent, the polymer-solvent interactions are overcome by those formed between chains, so random coil follows a tight and contracted conformation. In extremely "poor" solvents, the forces between polymer-solvent are eliminated thoroughly, and the random coil become so contracted that it precipitates [27, 28].

External factors, like agitation, temperature, and radiation exposure have a great impact on the dissolution process. It was found that the velocity of dissolution is higher as the agitation and stirring frequency of the solvent are intensified. This is because of the decrease of the thickness of the surface layer, while the dissolution rate reaches a critical value when the pressure of the solvent on the polymer surface is high, regardless the temperature [24]. Other reports [29–31] showed that in the absence of agitation the solvent diffuses into the polymer forming the gel layer, while subjection the solution to agitation the gel is stripped off very fast during the stirring stage. If in the first case the gel layer is diminished with time owing to desorption of the macromolecules, in the second case agitation favor sorption of solvent and subsequently desorption of chains from the swollen gel. Radiation dose is another factor that influences the dissolution process [32]. First, the polymer swells and then the chains begin to disentangle and dissolve. The swelling rate was higher in comparison to the dissolution rate, the gel layer size increased linearly with the square root of time. Oppositely, if the dissolution rate was bigger than the swelling one, then the gel thickness shrinks with time.

The polymer concentration in the solution has a strong impact on the overall material behavior viewed in the changes recorded in the viscosity (η). The Debye criterion, $c[\eta] = 1$ is useful to delimitate the concentration ranges of dissolute macromolecules. So, when discussing polymer solutions, there are three concentration domains [33]:

- The range of *dilute* solutions at $c[\eta] < 1$, where the macromolecules are mainly surrounded by solvent molecules so they behave independently of one another;
- The range of *semi-dilute* or *semi-concentrated* solutions at $1 < c[\eta] < 7$, where concentration is high enough that facilitates the contacts between different macromolecules, but there is no continuous three-dimensional entanglement network; and
- The range of *concentrated solutions* at $c[\eta] > 7$, where the macromolecular coils begin to overlap and thus an entanglement network is formed throughout the entire volume of solution.

It is presumed that polymers in dilute solution are independently moving and they do not display permanent contacts with each other [34]. As a result, viscosity of the dilute polymer solution is decreasing as the shearing is more intense reflecting a non-Newtonian behavior. This can be explained by deformation of the individual macromolecules in the stream. Polymer chains subjected to shearing modify their shape, thus reducing their resistance to flow owing to streamlining at higher shear rates [33]. On the other hand, for long-chain and flexible polymers the viscosity is unchanged at very low shear rates. This is the so-called the zero-shear rate Newtonian viscosity or limiting viscosity at null shear rate. The viscosity might further remain constant even at intermediate shear rates and at a critical shear rate the viscosity begin to decrease and at very higher shear rates reaches a minimum value, which is known as the upper Newtonian viscosity (or limiting viscosity) [33].

As the polymer amount in solution is higher more contacts between macromolecular coils occur and thus network of entanglements is formed [33, 35]. Chains at the sites of contact are able to slide and for this reason the network is considered quasi-permanent. The intermolecular contacts responsible for the occurrence of the entanglement network are presumed to appear at a critical concentration (noted c_e) as a function of the molecular weight. Non-Newtonian flow becomes more and more obvious with the augmentation of polymer concentration in solution [33]. Below the c_e point, the solution viscosity is strongly dependent on the intramolecular, excluded-volume effects, whereas above c_e the intermolecular entanglements have a prevalent impact on the rheology of the solution [33–35].

Most polymer products are obtained from semi-diluted or concentrated solutions [19, 36]. In these concentration domains, the analysis of the shear flow behavior is essential for proper processing/shaping of the polymer material. The rheology is the most suitable methods to examine the polymer fluid response to shear deformation [33]. The viscosity curves contain data of paramount importance concerning the flow-ability of polymer solution subjected to a broad interval of shear rates and processing conditions. The experiments recorded at small shear rates are especially useful for analysis of the problems which arise from manufacturing procedure. In the low shearing domain, the experiments are framed in linear viscoelastic region, and hence the data can be connected to the polymer's molecular structure [37–40]. The latter influences the polymer processability and the performance of the final product. This is why electronic, cosmetic or paint industry often employs rheological testing for product perfecting, as well as for novel product development.

Polymer films are generally prepared from solution phase using techniques like:

- **Drop Casting [41]:** Relies on preparation of a thin solid film by dropping a polymer solution onto a support with flat surface and left to evaporate the solvent;
- **Tape Casting [42]:** It is based on a reservoir of polymer solution with a slit-shaped outlet which leaves a layer of wet film;
- **Spin Coating Deposition [43]:** Small quantity of polymer solution is applied on the center of the substrate, which is subjected to spinning at variable rotation speeds or sometimes not spinning at all. In order to spread the solution, the substrate is rotated at high speed and thus the coating material is under action of centrifugal force and covers the support. Rotation is continued while the fluid reaches the edges of the support or until the wanted film is attained. The used solvent is generally volatile and mainly evaporates during spinning. As the rotation speed is higher, the obtained film is thinner [44];
- **Spray Deposition [45]:** It is a relatively new method that involves a highly diluted solution of a polymer, which is nebulized into air and concentrated under specific evaporation conditions. The

produced aerosol is transported by a carrier gas and cast onto a solid substrate;

- **Langmuir-Blodgett [46]:** It is mainly a nanostructured system obtained when Langmuir films—are transferred from the liquid-gas interface to solid substrate during the vertical crossing of the support through the monolayers.

Among the presented techniques of polymer film preparation from solution, phase the most commonly used ones are drop-casting, tape casting, and spin coating deposition. In these cases, flow behavior of the polymer solution is very important. The uniformity of the wet and implicitly solid film thickness is strongly related to the viscosity dependence of the shear rate. It was proved that Newtonian polymer fluids are suitable for obtaining a uniform liquid surface contour [47]. This is not the case of non-Newtonian solutions for which the thickness profile is far from uniform [48]. Particularly the polymer wet layer is thinner towards the substrate edge since solution viscosity is changing from wafer center to its margins as a result of a variation of the shear rate along spinning disk radius. This is also available for tape casting where the film thickness is affected by speed of blade movement and polymer solution response to variable shearing [42, 49]. Hence, the parameters extracted from rheological properties, such as the energy of activation and flow activation entropy, show the effect of the polymer structure, and provide more precise information on the proper processing conditions adapted to each macromolecular system. Having all these in view, a close understanding of the rheological functions (dynamic viscosity, shear stress, elastic shear modulus, loss tangent, viscous shear modulus, and relaxation time) of the polymer solutions and their dependence on temperature and solvent are of paramount importance for processing them into stable films with uniform thickness. The uniform film thickness is highly desirable for achieving isotropic physical properties (e.g., transparency or electrical/thermal conductivity) and securing the functionality of the components implemented in the devices.

1.3 FLOW BEHAVIOR OF SOME SYSTEMS CONTAINING GREEN POLYMERS AND THEIR APPLICABILITY

The rheological behavior of solutions prepared from green polymers as main or secondary components has known an increasing attention. Such materials can be viewed as additives for a large product portfolio, such as plasticizers, lubricants, viscosity depressants, antistatic, and antifogging components, but also as release, agents attained from sustainable materials that are prepared in such way that they optimize production and enhance efficiency [50]. In this section of the chapter, the state-of-art concerning rheological investigation of solutions containing certain green polymers is presented. The attention was directed towards the following types of green polymers:

- cellulose and its derivatives;
- silk;
- chitin; and
- chitosan.

1.3.1   SOLUTION RHEOLOGY OF CELLULOSE AND ITS DERIVATIVES

As already known, *cellulose* as found in nature is not soluble in water or in many other organic solvents because its chain stiffness and numerous inter and intramolecular interactions. However, cellulose dissolution was proved to be optimal in ionic liquids [51–55]. Most of them, particularly those halide anions are highly viscous fluids, having a viscosity of about three times higher in comparison with that of common organic solvents [51]. It was shown that even at elevated temperatures cellulose solution in chloride-based ionic liquids present large viscosity similar to certain polyolefin melts [54]. In practice, the high viscosity of cellulose-ionic liquids solutions is not so desirable because it will make difficult the mass transfer and require considerable power consumption for blending to achieve a heterogeneous solid-liquid system. The introduction of less viscous co-solvents into such solutions is a good alternative to reduce the mixture viscosity and enhance the efficiency of polymer processing, including heat and mass transfer rate [51]. For instance, dimethylsulfoxide, dimethylformamide, and pyridine have demonstrated to display a good ability to diminish the viscosity of cellulose solution in ionic liquids without an impact on the ionic liquids dissolving capability of polymer in a specific concentration domain of these co-solvents [56–58]. The addition of organic solvent is useful for adequate processing of cellulose-ionic liquids solutions with reduced costs. However, so far, literature is not abundant in studies concerning the viscoelastic behavior and solution behavior of cellulose solved in ionic liquids in the presence of co-solvents.

Lv and co-workers [51] studied the properties in solution of cellulose in a mixture of ionic liquid/dimethylsulfoxide. The viscosity is diminished by addition of dimethylsulfoxide in the system. It is presumed that this solved reduces monomer friction coefficient in the polymer solution, but it does not significantly modify the entanglement state of cellulose, while it shrinks the relaxation time. The steady shear experiments are useful to monitor the behavior of cellulose in dilute, semidilute unentangled and semidilute entangled domains. The investigated cellulose solutions display as Newtonian properties at very low polymer concentration, whereas the solution viscosity increases with concentration and presents a shear-thinning behavior at higher shear rates. The critical concentration for macromolecular coil overlapping (c*) and the entanglement concentration (c_e) of cellulose are found to be 0.4 wt% and 1.3 wt%, respectively. The parameters extracted from the power low concentration dependencies indicate that the ionic liquid/dimethylsulfoxide are like theta-solvents for cellulose at room temperature, while thermodynamic characteristics of the solvent mixtures were close to those of the ionic liquids for cellulose. In other words, the conformation of cellulose in ionic liquids would not be modified with the introduction of dimethylsulfoxide, not only in the dilute domain, but also in the entanglement concentration interval. These conclusions were supported by results achieved previously by Gericke et al. [52] for several types of cellulose in 1-ethyl-3-methylimidazolium acetate (EmimAc), proving that this ionic liquid is like a theta-solvent for cellulose. Sescousse and collaborators [53] performed a comprehensive investigation of viscosity concentration dependency in two distinct ionic liquids (EmimAc and 1-butyl-3-methylimidazolium chloride (BmimCl)) at several temperatures. They suggested that solvents present similar thermodynamic quality, in spite of their different state at room temperature, namely EmimAc is a liquid, and BmimCl is a solid. Chen et al. [54] studied linear viscoelastic and steady shear viscosity of cellulose solutions in BmimCl from dilute to highly entangled solutions. They obtained concentration dependencies of rheological functions consistent with scaling predictions for neutral polymers in a theta-solvent. However, failure of the Cox-Merz rule with the steady shear viscosity bigger than complex viscosity and the noticed internal mode structure of dilute and semidilute unentangled examined solutions indicate that cellulose in BmimCl is not simply a flexible polymer in a theta-solvent. Druel and co-workers [59] used JetCutting technology to prepare cellulose aerogel beads from ionic liquid solutions and dried them by supercritical carbon dioxide extraction. The rheological characteristics of cellulose solutions in 1,5-diazabicyclo[4.3.0]non-5-enium propionate ([DBNH][CO$_2$Et]) and in 1-ethyl-3-methylimidazolium acetate ([Emim][OAc]) were analyzed at several concentrations and temperatures. The first ionic liquid is a better solvent from thermodynamic point of view since the cellulose intrinsic viscosity is twice bigger in regard with that in [Emim][OAc]. From rheological data it was extracted the processing window of JetCutter which imposes a low solution viscosity at elevated shearing rates and also the polymer concentration above which the coils are overlapping ensuring the formation of intact aerogel beads. For their preparation, cellulose-[DBNH][CO$_2$Et] solutions having a concentration of 2–3 wt% were achieved and then coagulated in non-solvents like water and alcohols. Bead dimensions are varying from 0.5 to 0.7 mm when obtained from 2% solutions and up to 1.8 mm when made from 3% solution. The cellulosic beads have wide applicability in cosmetics, food, medical, sorption, filtration, and separation [59].

Owing to its insolubility, cellulose was chemically modified resulting a broad category of *cellulosic materials* that are more facile to be processed from solution phase. A recent study of Gauche and Felisberti [56] discusses the viscoelastic behavior of cellulose nanocrystals (CNCs) grafted with poly(2-alkyl-2-oxazoline)s. This procedure is a good strategy to enhance CNC dispersion in solvents (of low polarity) or polymeric matrixes. The grafting of cellulose with flexible macromolecules generates important changes in colloidal behavior. Poly(2-alkyl-2-oxazoline)s achieved by cationic ring-opening polymerization reaction was grafted onto CNC surfaces, where the coupling reaction was enabled by the presence of partially hydrolyzed polymers-attaining a reaction yield of about 64%. The modified cellulosic particles can be redispersed in water after freeze-drying, forming stable dispersions. In addition, they were not cytotoxic up to 10 wt% aqueous dispersion. The rheological properties of grafted nanoparticles indicate that there are significant differences in regard with original CNC dispersions. CNC dispersions in water present an organization typical for nematic liquid crystals and rheological behavior similar to true gel (at 5 wt%) before drying. Conversely, when in the same conditions is inserted 10 wt% of CNC-g-poly(2-ethyl-2-oxazoline-stat-ethylene imine) (P(EtOx-s-Ei)) the gel properties are less pronounced. Dispersions of CNC-g-P(PEtOx-s-Ei) particles prepared by redispersion of freeze-dried particles behaved as an isotropic fluid, lacking the nematic molecular order. For pristine CNC dispersions, the magnitudes of shear moduli were found to be almost independent of frequency having a prevalent elastic behavior. Moreover, the ratio of loss and elastic moduli has subunitary values in the range of applied angular frequency supporting the idea that the sample is a true gel. Upon addition of CNC-g-P(EtOx95-s-Ei5) to the initial CNC dispersion it is noticed that elastic modulus and viscosity decrease with 3–4 orders of magnitude, while frequency increased to 0.03 Hz, a shear-thickening zone was remarked. This could be connected to the coil-to-stretch transition in the tethered polymer chains subjected oscillatory shearing, wherein the macromolecule disentangles to an almost fully extended state. After 0.03 Hz, a shear-thinning zone appears since the effect of chain rigidity on the viscosity is no longer prevalent over the shearing, thus the alignment of the crystals along the flow direction is taking place, followed by breaking off the nematic organization [60]. The loss tangent (tanδ) values showed relevant modifications over the angular frequency interval: the transition from liquid-like to solid-like behavior was noticed at the overlap frequency, corresponding to tanδ≅ 1. This sol−gel transition appeared at the exact point that coincides with the beginning of the shear-thinning zone. In the range of low frequency, the tanδ value was 5.75, as expected for a viscous polymer solution. For higher frequencies, tanδ decreased up to a value of 0.17, and the CNC dispersion in the presence of 10 wt% CNC-g-POx might be regarded as a weak gel. These results support the hypothesis that the nanoparticle self-assembly was not present in these samples. A CNC dispersion attained from a physical blend of pristine CNCs and CNC-g-P(EtOx95-s-Ei5) followed by freeze-drying and redispersion in water (5 wt%) displayed a considerable reduction of the complex viscosity, shear moduli of several orders of magnitude. The CNC-g-P(EtOx95-s-Ei5) dispersion had a similar behavior.

Other cellulose derivatives, like hydroxypropyl cellulose (HPC) or hydroxypropylmethylcellulose (HPMS), have a rheological behavior strongly dependent on the solvent and concentration [37, 39, 61 64]. HPC forms in water, acetic acid or N,N-dimethylacetamide isotropic solutions at concentrations below 27%-value dependent on solvent and molecular weight. After that there is a small concentration interval where a biphasic (isotropic and cholesteric phases) behavior is remarked, followed by a cholesteric phase, where viscosity versus concentration has a peak. The anisotropic solutions of HPC are prevalently viscoelastic [61]. Upon heating, anisotropic HPC solutions in acetic acid endure an anisotropic to isotropic phase transition which is reflected in a maximum in the temperature variation of the first normal-stress difference and of the solution viscosity [61]. After shear cessation, HPC solutions exhibit a band texture-noticed as alternating bright and dark lines orthogonal to the shear direction caused by reorientation of the average molecular orientation. Such morphology is highly stable and is noticed when the macromolecules are strongly oriented along the shear direction [62–64]. The band texture can be maintained even after solution drying. Such surface pattern can be transferred upon blending to other polymers with isotropic orientation of macromolecular chains [37, 39, 65]. Cellulose derivative films with banded pattern on their surface can be used for cell culture purposes or for nematic, alignment in liquid crystal displays [37].

1.3.2   SOLUTION RHEOLOGY OF SILK POLYMERS

The study of Keerl [66] was focused on rheological characterization of regenerated *Bombyx mori* fibroin and silk solutions in water of recombinant spider silk protein. The introduction of kosmotropic salts and application of elevated temperatures determine a solution behavior similar to that of a native silk spinning dope. In the studied shear rate interval, the silk-based samples presented a non-Newtonian behavior. This is expected because of the increasing flow alignment of the underlying microstructure. The onset of the shear thinning domain, that is, the critical shear rate noticed in other reports on natural silk dopes, was usually around 0.5–4 s^{-1} [67–69]. Measurements performed on the silk solutions highlighted shear-thinning right from the beginning, with critical shear rates starting around 0.01 s^{-1}. Oscillatory tests below 5% w/v show a storage modulus independent of frequency viewed as a plateau between shear frequencies of 2 and 10 rad/s [66] indicating a gel-like behavior. Above 5% w/v, silk solutions in water present frequency dependence and an overlap point about 15–30 rad/s, which is in good accordance with the viscoelastic properties of native silk dope [66]. These analyses are relevant for determining proper conditions of achieving green polymer fibers of elevated mechanical resistance used in various domains, from biomedicine to military purposes [66].

The investigation of Tao and collaborators [70] was focused on flow behavior of silk fibroin (SF) reinforced waterborne polyurethane (WPU). These green materials have promising features for tissue engineering (TE) applications. In the shear rate domain of $10^{-2}–10^3$ s^{-1}, the composite solutions have a shear thinning behavior regardless the amount of WPU in the system. This is sustained by the subunitary values of the flow index extracted from Ostwald-de Waele law applied to shear stress data as a function of shear rate. Dynamic oscillation behavior of the SF-WPU solutions. All WPU dispersions were found to be highly frequency dependent. At 13 wt%, shear moduli curves were crossed at the middle of the frequency range, displaying a rheological behavior similar to a dilute solution. When the percent of WPU dispersion is varying between 15 to 25 wt%, storage modulus is lower than that of the loss one and parallels to each other at low frequency range, while the moduli approach each other at high shear frequencies. This proves that the dispersions have features analogs to Brownian suspensions. At high WPU contents, like 22 and 25 wt% one may notice a gel-like network structure or aggregates formation, which is seen in higher viscosity compared to that of dispersions containing lower WPU amounts. The formation of aggregates or gel-like networks in dispersion system with WPU contents above 17 wt% might restrict the dispersion of SF particles into the polymer matrix.

Laity et al. [71] analyzed the rheological features of native silk protein feedstock specimens at temperatures ranging from 2 to 55°C. They observed no thermally-driven phase change behavior. All solutions have flow properties commonly noted for concentrated polymer solution, with viscoelastic characteristics dominated by two main relaxation modes with time constants around 0.44 and 0.055 s at 25°C. The samples present a temperature dependence following the Arrhenius relation, consistent with the kinetics which is governed by activation energy of flow, which varied from 30.9 to 55.4 kJ mol⁻¹ according to oscillatory data. The compiled master-curves for native silk feedstock samples based on the principles of time-temperature superposition, using oscillatory data prove their viscoelastic behavior peculiar for a polymer solution across a wide temperature range. The rheological data showed the processing range of natural silks and explained the molecular mechanisms governing the flow properties of these green materials.

1.3.3   SOLUTION RHEOLOGY OF CHITIN

Bochek and collaborators [72] reported the rheological behavior of cellulose/chitin mixtures in 1-butyl-3-methylimidazolium acetate (BMIAc). They noticed a non-Newtonian flow of this green polymer blend in solution. Positive and negative deviations of values of the viscosity from additive values are noted even if polymerization degree of chitin is higher than that of cellulose. This is probably because destruction of polysaccharide chains takes place during dissolution in BMIAc, but that of chitin is more pronounced [72]. Another reason for this result arises from the fact that BMIAc is a thermodynamically lower solvent for cellulose and better for chitin. Addition of more chitin in the system determines a decrease of the solution viscosity. The network of intermolecular entanglements and the hydrogen bonds often noted for cellulose are disrupted, resulting a new blend system of hydrogen bonds, concomitantly with a weaker network of intermolecular entanglements. Such aspects determine a lower energetic barrier of flow process.

1.3.4 SOLUTION RHEOLOGY OF CHITOSAN

The rheological properties of chitosan solutions in several solvents-acetic, lactic, and hydrochloric acid-and various temperatures (25–45°C) were reported [73]. At shear rates up to 1 s⁻¹, the real part of the complex viscosity is almost Newtonian after which it decreases. The analyzed chitosan solutions present a loss modulus higher than the elastic one showing that viscous effects are more significant than elastic ones. Application of the Cox-Merz rule for 2% chitosan solution in acetic acid indicates a decrease of the viscosity and thus a lower energetic barrier of flow.

Hydrophobically modified chitosan was prepared by Shedge and Badgier [74] from natural resources, like cashew nut shell liquid. Chemical modification of this green polymer was done with 3-pentadecyl cyclohexane carbaldehyde. Above the critical association concentration, the hydrophobically modified chitosan solutions present aggregation properties. As the polymer concentration is higher, the intermolecular hydrophobic associations begin to play a prevalent role and determine the formation of flower-like micelles. The frequency sweep experiments beyond the overlap concentration reveal for elastic modulus weak frequency dependence, having almost a plateau modulus at about 1 Pa, while loss modulus increases with frequency. The increase of hydrophobic content determines the increase of shear moduli and the disappearance of their overlap, showing a soft-solid gel-like behavior. For the same degree of hydrophobic modification and variable concentration, it was found that there is an evolution from liquid-like behavior to a solid-like structure, which reveals a transition to a gel phase. For the latter case, the amount of elastically active chains, that is, the bridges between flower-like micelles is higher and form an extensive network which is reflected in increased viscosity of solutions. These materials have a great potential in applications like thickeners for cosmetic lotions, paints, oil recovery, textile materials, and other industrial products.

Chitosan/glycerol-phosphate solutions present a thermo-gelling behavior as reported by Chenite et al. [75]. They prove that chitosan solutions can be neutralized and no gel-like precipitation is taking place upon using glycerol-phosphate. Subjecting the samples to increased temperatures it was noticed the formation of hydrogel. The addition of the basic salt modulates hydrophobic interactions and hydrogen bonding which are useful to keep chitosan in solution at neutral pH at 4°C, and then allow thermogelling at 37°C. Rheological data reveal endothermic gelation of samples and facilitated the formulation of a sol/gel diagram. The gel occurrence is the result of the interplay between temperature and pH. These materials are utilized in pharmaceutical and food industries.

Lewandowska [76] analyzed the rheological characteristics of chitosan blends with partially hydrolyzed polyacrylamide (HPAM) in several solvents (aqueous acetic acid, lactic acid (LA), and aqueous acetic acid/NaCl). The examined chitosan-based solutions displayed a non-Newtonian behavior with shear-thinning and/or shear-thickening zones. In other words, the flow index values are subunitary as expected for pseudoplastic behavior for HPAM, chitosan, and their blends in most solvents. When using aqueous acetic acid, the flow index is higher than 1 showing shear-thickening behavior.

The highest flow activation energy was observed for the polymer solutions in aqueous acetic acid. In addition, this parameter is higher for the polymer blends in regard with that of the individual counterpart's solutions in aqueous 0.1 M CH$_3$COOH/0.2 M NaCl. The result reveals that there are certain interactions among the polymer components in the blend systems.

1.4 CONCLUDING REMARKS

The market of green polymer materials is in continuous progress as the environmental requirements are demanding less polluting products. The rheological behavior of polymers is paramount for the processing stage. Polymer films of uniform thickness are achieved when the solution has a viscosity independent of the shear deformation. The role of rheology in solution processing into polymer films is discussed. Several systems of green polymers in solution are reviewed from rheological point of view showing that the solvent, temperature, or additive is strongly impacting the viscosity dependence on shear rate and viscoelastic behavior.

ACKNOWLEDGMENT

The financial support of European Social Fund for Regional Development, Competitiveness Operational Programme Axis 1-Project "Petru Poni Institute of Macromolecular Chemistry-Interdisciplinary Pol for Smart Specialization through Research and Innovation and Technology Transfer in Bio(nano)polymeric Materials and (Eco)Technology," InoMatPol (ID P_36_570, Contract 142/10.10.2016, cod MySMIS: 107464) is gratefully acknowledged.

KEYWORDS

- cellulose nanocrystals
- green polymers
- processability
- viscoelasticity
- viscosity
- waterborne polyurethane

REFERENCES

1. Ellis, B., (2000). Environmental issues in electronics manufacturing: A review. *Circuit World, 26,* 17–21.
2. *How Industrial Applications in Green Chemistry are Changing Our World.* https://chemical.report/whitepapers/how-industrial-applications-in-green-chemistry-are-changing-our-world/2067 (accessed on 3 August 2020).
3. Doble, M., & Kruthiventi, A. K., (2007). *Catalysis and Green Chemistry, in Green Chemistry and Engineering* (pp. 53–67). Elsevier: India.
4. Anastas, P., (2011). Twenty years of green chemistry. *Chem. Eng. News, 89*(26), 62–65.
5. *Green and Natural Polymers are on the Rise.* https://www.polymersolutions.com/blog/green-and-natural-polymers-on-the-rise/ (accessed on 3 August 2020).
6. Jin, S., Byrne, F., McElroy, C. R., Sherwood, J., Clark, J. H., & Hunt, A. J., (2017). Challenges in the development of bio-based solvents: A case study on methyl(2,2-dimethyl-1,3-dioxolan-4-yl)methyl carbonate as an alternative aprotic solvent. *Faraday Discuss, 202,* 157–173.
7. Kobayashi, S., (2017). Green polymer chemistry: New methods of polymer synthesis using renewable starting materials. *Struct. Chem.,28,* 461–474.
8. Worthington, M. J. H., Kucera, R. L., & Chalker, J. M., (2017). Green chemistry and polymers made from sulfur. *Green Chem., 19,* 2748–2761.
9. Gomez, J. G. C., Méndez, B. S., Nikel, P. I., Pettinari, M. J., Prieto, M. A., & Silva, L. F., (2011). Making green polymers even greener: Towards sustainable production of polyhydroxyalkanoates from agro-industrial by-products. In: Petre, M., (ed.), *Advances in Applied Biotechnology* (pp. 41–62). InTech: Croatia.
10. Hong, M., & Chen, E. Y. X., (2019). Future directions for sustainable polymers. *Science and Society, Special Issue Part Two: Big Questions in Chemistry, 1,* 148–151.

11. Gawel, E., Pannicke, N., & Hagemann, N., (2019). A path transition towards a bio-economy-The crucial role of sustainability. *Sustainability, 11*, 3005.

12. Thompson, R. C., Moore, C. J., Vom, S. F. S., & Swan, S. H., (2009). Plastics, the environment and human health: Current consensus and future trends. *Philos Trans. R. Soc. Lond. B Biol. Sci., 364*, 2153–2166.

13. Prata, J. C., Silva, A. L. P., Da Costa, J. P., Mouneyrac, C., Walker, T. R., Duarte, A. C., & Rocha-Santos, T., (2019). Solutions and integrated strategies for the control and mitigation of plastic and microplastic pollution. *Int. J. Environ. Res. Public Health, 16*, 2411.

14. D'Ambrières, W., (2019). Plastics recycling worldwide: Current overview and desirable changes. *Field Actions Science Reports* [Online], p. 19 http://journals.openedition.org/factsreports/5102 (accessed on 3 August 2020).

15. Mathers, R. T., & Meier, M. A. R., (2011). *Green Polymerization Methods: Renewable Starting Materials, Catalysis, and Waste Reduction.* Wiley: Germany.

16. National Research Council, (1994). *Polymer Science and Engineering: The Shifting Research Frontiers.* Washington, DC: The National Academies Press. https://doi.org/10.17226/2307 (accessed on 3 August 2020).

17. Denson, C. D., (1984). Polymer processing operations other than shaping. In: Astarita, G., & Nicolais, L., (eds.), *Polymer Processing and Properties.* Springer: Boston.

18. Baird, D. G., & Collias, D. I., (2014). *Polymer Processing: Principles and Design.* Wiley: USA.

19. Han, C. D., (2007). Rheology and processing of polymeric materials. *Volume 1: Polymer Rheology.* Oxford University Press: USA.

20. Miller-Chou, B. A., & Koenig, J. L., (2003). A review of polymer dissolution. *Prog. Polym. Sci., 28*, 1223–1270.

21. Su, W. F., (2013). Polymer size and polymer solutions. In: Su, W. F., (ed.), *Principles of Polymer Design and Synthesis* (pp. 9–26). Lecture notes in chemistry. Springer: Germany.

22. Danner, R. P., & High, M. S., (1993). *Handbook of Polymer Solution Thermodynamics.* Wiley: USA.

23. Tsuchida, E., (1982). *Interactions between Macromolecules in Solution and Intermacromolecular Complexes.* Springer: USA.

24. Ueberreiter, K., (1968). The solution process. In: Crank, J., & Park, G. S., (eds.), *Diffusion in Polymers* (pp. 219–257). Academic Press: New York.

25. Cooper, W. J., Krasicky, P. D., & Rodriguez, F., (1985). Effects of molecular weight and plasticization on dissolution rates of thin polymer films. *Polymer, 26*, 1069–1072.

26. Manjkow, J., Papanu, J. S., Hess, D. W., Soane (Soong), D. S., & Bell, A. T., (1987). Influence of processing and molecular parameters on the dissolution rate of poly-(methyl methacrylate) thin films. *J. Electrochem. Soc., 134*, 2003–2007.

27. Wang, R., & Wang, Z. G., (2012). Theory of polymers in poor solvent: Phase equilibrium and nucleation behavior. *Macromolecules, 45*, 6266–6271.

28. Ouano, A. C., & Carothers, F. A., (1980). Dissolution dynamics of some polymers: Solvent-polymer boundaries. *Polym. Eng. Sci., 20*, 160–166.

29. Pekcan, O., Ugur, S., & Yilmaz, Y., (1997). Real-time monitoring of swelling and dissolution of poly(methyl methacrylate) discs using fluorescence probes. *Polymer, 38*, 2183–2189.

30. Pekcan, O., & Ugur, S., (2002). Molecular weight effect on polymer dissolution: A steady state fluorescence study. *Polymer, 43*, 1937–1941.
31. Ugur, S., & Pekcan, O., (2000). Fluorescence technique to study thickness effect on dissolution of latex films. *J. Appl. Polym. Sci., 77*, 1087–1095.
32. Drummond, R., Boydston, G. L., & Peppas, N. A., (1990). Properties of positive resists. III. The dissolution behavior of poly(methyl methacrylate-co-maleic anhydride). *J. Appl. Polym. Sci., 39*, 2267–2277.
33. Malkin, A. Y., & Isayev, A. I., (2006). *Rheology: Concepts, Methods, and Applications.* ChemTec. Publishing: Toronto.
34. *How Polymers Behave in Dilute Solutions.* https://pslc.ws/macrog/property/solpol/ps5. htm (accessed on 3 August 2020).
35. Chisca, S., Barzic, A. I., Sava, I., Olaru, N., & Bruma, M., (2012). Morphological and rheological insights on polyimide chain entanglements for electro spinning produced fibers. *J. Phys. Chem. B., 116*, 9082–9088.
36. Han, C. D., (2007). Rheology and processing of polymeric materials. *Volume 2: Polymer Processing.* Oxford University Press: USA.
37. Cosutchi, A. I., Hulubei, C., Stoica, I., & Ioan, S., (2010). Morphological and structural-rheological relationship in epiclon-based polyimide/hydroxypropylcellulose blend systems. *J. Polym. Res., 17*, 541–550.
38. Barzic, A. I., Rusu, R. D., Stoica, I., & Damaceanu, M. D., (2014). Chain flexibility versus molecular entanglement response to rubbing deformation in designing poly(oxadiazolenaphthylimide)s as liquid crystal orientation layers. *J. Mater. Sci., 49*, 3080–3098.
39. Stoica, I., Barzic, A. I., & Hulubei, C., (2016). Polyimide embedding in lyotropic polymer matrix for surface-related applications: Rheological and microscopy investigations. *Rev. Roum. Chim., 61*, 575–581.
40. Cosutchi, A. I., Hulubei, C., & Ioan, S., (2007). Rheological study of some epiclon-based polyimides. *J. Macromol. Sci. Part B., 46*, 1003–1012.
41. Zhao, K., Yu, X., Li, R., Amassian, A., & Han, Y., (2015). Solvent-dependent self-assembly and ordering in slow-drying drop-cast conjugated polymer films. *J. Mater. Chem. C., 3*, 9842–9848.
42. Albu, R. M., Hulubei, C., Stoica, I., & Barzic, A. I., (2019). Semi-alicyclic polyimides as potential membrane oxygenators: Rheological implications on film processing, morphology, and blood compatibility. *Express Polym. Lett., 13*, 349–364.
43. Barzic, A. I., Soroceanu, M., Albu, R. M., Ioanid, E. G., Sacarescu, L., & Harabagiu, V., (2019). Correlation between shear-flow rheology and solution spreading during spin coating of polysilane solutions. *Macromol. Res*. doi: 10.1007/s13233-019-7164-7.
44. Bornside, D. E., Macosko, C. W., & Scriven, L. E., (1989). Spin coating: One-dimensional model. *J. Appl. Phys., 66*, 5185–5193.
45. Fujita, K., Ishikawa, T., & Tsutsui, T., (2002). Novel method for polymer thin film preparation: Spray deposition of highly diluted polymer solutions. *Jpn. J. Appl. Phys., 41*, L70.
46. Matsui, J., Yoshida, S., Mikayama, T., Aoki, A., & Miyashita, T., (2005). Fabrication of polymer Langmuir-Blodgett films containing regioregular poly(3-hexylthiophene) for application to field-effect transistor. *Langmuir, 21*, 5343–5348.
47. Emslie, A. G., Bonnet, E. T., & Peck, L. G., (1958). Flow of a viscous liquid on a rotating disk. *J. Appl. Phys., 29*, 858–862.

48. Acrivos, A., Shah, M. J., & Petersen, E. E., (1960). On the flow of a non-Newtonian liquid on a rotating disk. *J. Appl. Phys., 31*, 963–968.
49. Berni, A., Mennig, M., & Schmidt, H., (2004). Doctor blade. In: Aegerter, M. A., & Mennig, M., (eds.), *Solgel Technologies for Glass Producers and Users* (pp. 89–92). Springer: Boston.
50. *Green Polymer Additives.* https://greenpolymeradditives.emeryoleo.com/additives/ (accessed on 3 August 2020).
51. Lv, Y., Wu, J., Zhang, J., Niu, Y., Liu, C. Y., He, J., & Zhang, J., (2012). Rheological properties of cellulose/ionic liquid/dimethylsulfoxide (DMSO) solutions. *Polymer, 53*, 2524–2531.
52. Gericke, M., Schlufter, K., Liebert, T., Heinze, T., & Budtova, T., (2009). Rheological properties of cellulose/ionic liquid solutions: From dilute to concentrated states. *Bio-Macromolecules, 10*, 1188–1194.
53. Sescousse, R., Le, K. A., Ries, M. E., & Budtova, T., (2010). Viscosity of cellulose-imidazolium-based ionic liquid solutions. *J. Phys. Chem. B., 114*, 7222–7228.
54. Chen, X., Zhang, Y. M., Wang, H. P., Wang, S. W., Liang, S. W., & Colby, R. H., (2011). Solution rheology of cellulose in 1-butyl-3-methyl imidazolium chloride. *J. Rheol., 55*, 485–494.
55. Druel, L., Niemeyer, P., Milow, B., & Budtova, T., (2018). Rheology of cellulose-[DBNH][CO$_2$Et] solutions and shaping into aerogel beads. *Green Chem., 20*, 3993–4002.
56. Fort, D. A., Swatloski, R. P., Moyna, P., Rogers, R. D., & Moyna, G., (2006). Use of ionic liquids in the study of fruit ripening by high-resolution ^{13}C NMR spectroscopy: 'Green' solvents meet green bananas. *Chem. Commun., 7*, 714–716.
57. Evlampieva, N. P., Vitz, J., Schubert, U. S., & Ryumtsev, E. I., (2009). Molecular solutions of cellulose in mixtures of ionic liquids with pyridine. *J. Appl. Chem., 82*, 666–672.
58. Rinaldi, R., (2011). Instantaneous dissolution of cellulose in organic electrolyte solutions. *Chem. Commun., 47*, 511–513.
59. Gauche, C., & Felisberti, M. I., (2019). Colloidal behavior of cellulose nano crystals grafted with poly(2-alkyl-2-oxazoline)s. *ACS Omega, 4*, 1189 3–11905.
60. Larson, R. G., & Magda, J. J., (1989). Coil-stretch transitions in mixed shear and extensional flows of dilute polymer solutions. *Macromolecules, 22*, 3004–3010.
61. Barzic, A. I., Albu, R. M., Gradinaru, L. M., & Buruiana, L. I., (2018). New insights on solvent implications in flow behavior and interfacial interactions of hydroxypropylmethylcellulose with cells/bacteria. *E-Polymers, 18*, 135–142.
62. Navard, P., & Haudin, J. M., (1986). Rheology of hydroxypropyl cellulose solutions. *J. Polym. Sci., 4*, 189–227.
63. Ernst, B., & Navard, P., (1989). Band textures in mesomorphic (hydroxypropyl) cellulose solutions. *Macromolecules, 22*, 1419–1422.
64. Peuvrel, E., & Navard, P., (1991). 24 Band textures of liquid crystalline polymers in elongational flows. *Macromolecules*, 5683–5686.
65. Barzic, A. I., Hulubei, C., Avadanei, M. I., Stoica, I., & Popovici, D., (2015). Polyimide precursor pattern induced by banded liquid crystal matrix: Effect of dianhydride moieties flexibility. *J. Mater. Sci., 50*, 1358–1369.
66. Keerl, D., & Scheibel, T., (2014). Rheological characterization of silk solutions. *Green Mater., 2*, 11–23.
67. Kojic, N., Bico, J., Clasen, C., & McKinley, G. H., (2006). *Ex vivo* rheology of spider silk. *J. Exp. Biol., 209*, 4355–4362.

68. Moriya, M., Roschzttardtz, F., Nakahara, Y., Saito, H., Masubuchi, Y., & Asakura, T., (2009). Rheological properties of native silk fibroins from domestic and wild silkworms, and flow analysis in each spinneret by a finite element method. *Biomacromolecules, 10,* 929–935.

69. Terry, A. E., Knight, D. P., Porter, D., & Vollrath, F., (2004). pH-induced changes in the rheology of silk fibroin solution from the middle division of *Bombyx mori* silkworm. *Biomacromolecules, 5,* 768–772.

70. Tao, Y., Hasan, A., Deeb, G., Hu, C., & Han, H., (2016). Rheological and mechanical behavior of silk fibroin reinforced waterborne polyurethane. *Polymers, 8,* 94.

71. Laity, P. R., & Holland, C., (2017). Thermo-rheological behavior of native silk feedstocks. *Eur. Polym. J., 87,* 519–534.

72. Bochek, A. M., Murav'ev, A. A., Novoselov, N. P., Zabivalova, N. M., Petrova, V. A., Yudin, V. E., Popova, E. N., & Lavrent'ev, V. K., (2013). Rheological properties of mixtures of cellulose with chitin in 1-butyl-3-methylimidazolium acetate of the obtained composite films obtained. *J. Appl. Chem., 86,* 1913–1917.

73. Martínez-Ruvalcaba, A., Chornet, E., & Rodrigue, D., (2004). Dynamic rheological properties of concentrated chitosan solutions. *Appl. Rheol., 14,* 140–147.

74. Shedge, A. S., & Badiger, M. V., (2014). Hydrophobically modified chitosan: Solution properties by rheology and light scattering. *Advancesin Planar Lipid Bilayers and Liposomes, 19,* 91–102.

75. Chenite, A., Buschmann, M., Wang, D., Chaput, C., & Kandani, N., (2001). Rheological characterization of thermo gelling chitosan/glycerol-phosphate solutions. *Carbohydr. Polym., 46,* 39–47.

76. Lewandowska, K., (2012). Rheological properties of chitosan blends with partially hydrolyzed polyacrylamide in different solvents. *Progr. Chem. Appl. Chitin and its Derivatives, 17,* 53–62.

CHAPTER 2

Properties of the Magnetic Polymer Nanocomposites in Magnetic Fields

JIMSHER ANELI[1] and GRIGOR MAMNIASHVILI[2]

[1]Institute of Machine Mechanics, 10, Mindeli Str., Tbilisi – 0186, Georgia, E-mail: jimaneli@yahoo.com

[2]Andronikashvili Institute of Physics of Ivane Javakhishvili Tbilisi State University, 6, Tamarashvili Str., Tbilisi – 0177, Georgia

ABSTRACT

The processes of self-assembly were studied in the magnetic polymer carbon nanocomposites doped with cobalt nanoclusters. These processes proceed due to the diffusion of magnetic nanoparticles stimulated by a combined effect of an outer steady magnetic fields and heating. The obtained polymer composites are promising for practical applications.

2.1 INTRODUCTION

In the last decade, the investigation of such new nanostructure forms of carbon as nanoparticles, nanotubes, and nanowires has become very topical. This is associated with the fact that, due to their sizes and peculiarities of their atomic structure, nanostructural particles reveal such unique physical-mechanical properties that the range of their promising applications covers many areas of activity from microelectronics to medicine.

In recent years, there increased interest in technologies of production of carbon-based materials oriented on the production of doped carbon nanoparticle modifications (nanotubes, nanoclusters, nanowires). This gives scientists and engineers the opportunity of aimed control of the unique properties of these materials, which are their natural properties [1].

As a matter of fact, the nanoobject control at the nanometer level using nanoparticles with the aim to arrange them in rows, signatures, and grids is the clue to the production of new functional materials. Hence, in recent years, for obtaining of constructional units of different nanometric sizes, many methods of self-assembling and synthesis were developed. In this connection, the possibility to control perfectly the self-assembling and synthesis processes of nanoparticles is a serious challenge from the point of view of both fundamental and applied investigations.

Based on the fundamental principles, the process of self-assembling requires the existence of an interaction between atoms and clusters, as well as of thermodynamic and kinetic driving forces, so that the organization of atoms and clusters for creation of nano-size domain structures should be realized. From this point of view, magnetic nanoparticles deserve a particular interest due to their unique physical-chemical properties and applicability in the new functional material technologies.

Carbon shells provide both the protection of ferromagnetic impurities against aggressive environments and new unique properties for the hydride nanostructures. The self-assembling of magnetic clusters coated with carbon shells represents just such an example that could be used in the contemporary materials, for instance, in strong rare-earth free bonded magnets, analytical instruments (nuclear magnetic resonance tomographs) and nanosensors using magnetic one-dimensional nanowires.

Moreover, currently, due to their low toxicity, magnetic carbon nanoparticles are under testing for therapeutic and diagnostic applications.

In recent years, the magnetic field was used for the creation of nanoscale materials, which resulted in significant achievements in fabrication of macro- and microstructure synthesized materials possessing unique properties.

In contrast with other existing self-assembling technologies, ordering induced by the magnetic field defines the formation of magnetic nanoparticles in ordered structures with unique properties. Therefore, the area of application of carbon magnetic nanoparticles is quite large. It is enough to name such applications as magnetic fluids, plastic scratch-resist glasses, information storage magnetic media, sensors, biomedicine, etc. It should be noted that, in spite of their broad prospects for multifunctional applications, the carbon nanoparticles doped with ferromagnetic clusters have not been well researched [2].

In work [3] the electric and magnetic properties measurements were carried out to study the gradiently anisotropic conducting and magnetic polymer composites synthesized due to self-assembly processes of nanoparticles under influence of the elastic forces developed by the stretching of polymer film composites fabricated on the basis of polyvinyl alcohol (PVA) doped with graphite powder and nickel nanoparticles.

In this work, the study of self-assembling properties of carbon nanoparticles doped with cobalt nanoclasters in magnetic polymer nanocomposites (PNC) under the combined influence of magnetic field and heating will be carried out using methods similar ones developed in work [3].

2.2 RESULTS AND DISCUSSION

To achieve this goal, we planned experiments with magnetic nanopowders, preparation of filled and unfilled polymer films, and the study of self-organization processes in them. This process was facilitated by heating the films above the glass transition temperature. The particles in conventional composites are essentially immobile in contrast to PNC particularly above the glass transition temperature T_g. The nanoparticles mobility can affect polymer dynamics resulting in changes in the viscosity modulus, kinetics of the particle-cluster formation, etc., [3].

The tensile measurements showed that, below T_g, the conventional composites and PNCs behave similarly with respect to mechanical properties. Though, above T_g, the toughness of PNC can increase by an order of magnitude with increasing temperature. It was assumed that the mechanism of toughness enhancement is the mobility of nanoparticles. The development of self-healing materials and coatings where nanoparticles migrate towards various defect sites requires better understanding of the process of nanoparticle diffusion. Heating of PNC above T_g enhances the mobility of polymer chains, which should facilitate the boundary diffusion between polymer interfaces, and this effect should be visualized using the magnetic nanoparticles introduced in the polymer. This process could be improved by applying additional stimuli, in particular a low-frequency (ac) magnetic field, a stationary magnetic field, pressure, heating separately or in combinations, etc. Such an impact stimulates self-assembling processes in the prepared films in the result of which one could produce the films "glued" to each other without using other type glues and polymer melting temperature. One of the objectives of this work was the development of a simple technology of production of carbon nanoparticles doped with ferromagnetic clusters and the study of their morphology and composition.

In particular, for production of carbon-based nanopowders and nano-coatings, the method of chemical vapor deposition (CVD) is used mainly along with application of the process of hydrogen reduction of volatile chlorides. The carbon nanoparticles doped with cobalt magnetic nanoclusters with mean sizes in the range 50–100 nm were synthesized by technology using the combination of pyrolysis of ethanol (and other hydrocarbons), vapor pyrolysis, and the CVD process in the mode of a closed recirculation de cycle with monitored technology parameters.

The developed technological process was realized in the installation, the reactor design and basic units of which provided for the possibility of monitoring of parameters such as the vapor content in reactor zones, catalytic capacity of substrates, partial oxygen pressure (over the range of $10^{-20} \div 10^{-25}$ atm.). This allows carrying out the investigations with the aim to establish the optimal technological parameters for production of finely dispersed carbon nanopowders doped with magnetic nanoclusters. Detailed description of this technology is given in work [4].

For preparation of polymer films, polyvinyl butyral (PVB) polymer with low $T_g \sim 45$–$55°C$ was chosen. Polymer PVB is a resin mostly used wherever strong binding, optical clarity, adhesion to many surfaces, toughness, and flexibility is required. As filler, we used the carbon nanopowder doped with magnetic (Co or Fe) nanoclusters (C/Co) of our production. For a comparative study, we used commercial Co nanopowder with average diameter of 28 nm (Sun Co., USA). The concentration of used Co-doped carbon nanopowders in the polymeric composite was in the range of 10–50 wt. %, while for the Co nanopowder filled polymer composite concentration was equal to 20 wt. %.

At first 10%, alcohol solution was prepared, than this solution was poured into Teflon press molds and, after their drying for 48 hours, the films one mm thick was obtained. The filled composites were prepared as follows: magnetic nanopowders were taken in the appropriate proportion (in terms of dry weight) and PVB was mixed with alcohol in a usual way, and then followed the ultrasonic treatment for 10–15 min for the destruction of magnetic nanopowder agglomerates. After thorough mixing for 7–10 min, magnetic polymer composite films similar to unfilled ones were obtained in Teflon press molds. From these films, the circular shape disk samples were cut.

To study the self-assembly processes in these PNC at different concentrations of carbon magnetic nanopowder, we used a simple method from work [5]. In this case, circular samples of the polymer composite (diameter-28 mm, thickness-1 mm) were exposed to the magnetic field which was provided by two attached permanent neodymium magnets and temperature of 85°C for two hours (Figure 2.1).

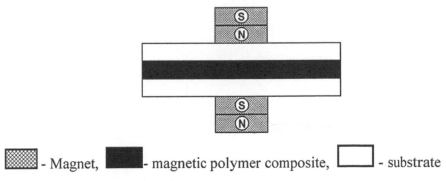

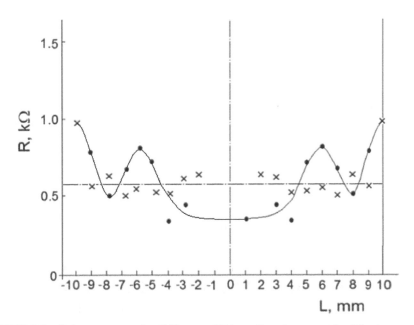

- Magnet,   - magnetic polymer composite,   - substrate

FIGURE 2.1   Geometry of the samples.

Resulting self-assembly of C/Co nanopowders caused changes in their concentration and modulation of local resistance along the radius of the sample, which was measured by a two-contact method as in our previous work [6] (Figures 2.2–2.7). The resistance was measured between the points spaced 2 mm apart along the radius in all following cases except for Figure 2.3 where the resistance was measured between the sample center and the points 2 mm.

FIGURE 2.2   Polymer composite C/Co, wt. 30%. ×: Sample prepared without magnetic field treatment; •: sample prepared under combined magnetic field and heat treatment.

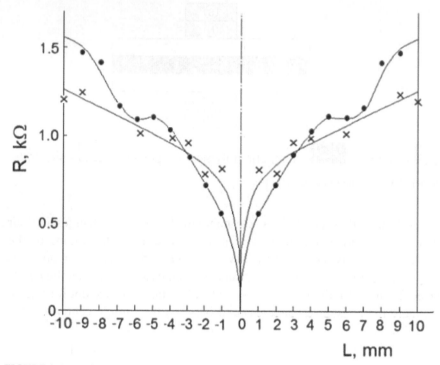

FIGURE 2.3 Polymer composite C/Co, wt. 30%.

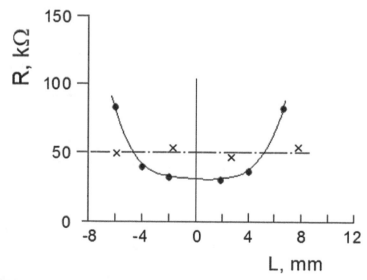

FIGURE 2.4 Polymer composite C/Co, wt. 50%

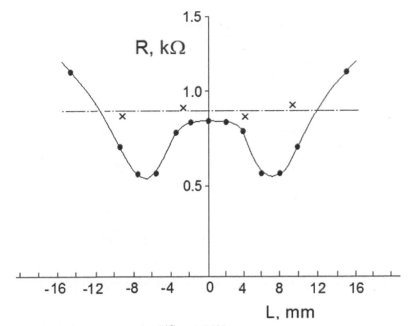

FIGURE 2.5   Polymer composite C/Co, wt. 20%.

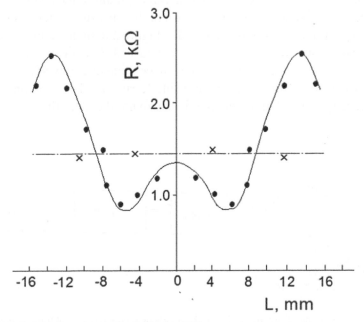

FIGURE 2.6   Polymer composite C/Co, wt. 15%.

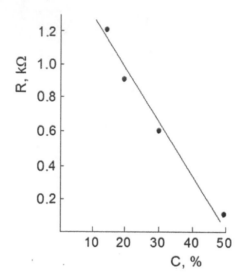

FIGURE 2.7 Dependence of resistance of initial untreated samples on the nanopowder concentration.

Because polymeric composites are magnetic carbon nanocomposites, we were able to study the processes of the self-assembling of nanoparticles using radio frequency resonance magnetic susceptibility measuring devices [5].

The experimental set-up is presented in Figure 2.8. In the inductive coil of the resonance contour of LC-generator, cylindrical tipped ferrite rod is used as a probe. The investigated rectangular shape magnetic polymer composite film is displaced relatively the immovable ferrite tip. The scanning of the film surface is realized along the previously marked disk radius.

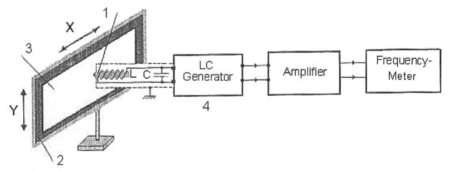

FIGURE 2.8 Scheme of the measuring of the magnetic characteristics of the polymer films. 1-ferrite probe; 2-frame; 3-magnetic polymer nanocomposite film; 4-LC generator.

The change of magnetic particle concentration causes the inductance change dL of the resonance contour of LC-generator resulting in the frequency displacement of LC-generator df related with dL by relation $df/f = \frac{1}{2}\, dL/L$. This frequency displacement could be precisely measured that stipulates the high sensitivity of the method. At the natural frequency of used LC-generator near ~ 2 MHz the observed range of the frequency change df was about ~ 1000 Hz at the precision of the frequency measurement ~ 1 Hz.

For example, we represent the result of self-assembly measurements of Co/C magnetic nanocomposite film in the magnetic field of neodymium magnet (Figure 2.9) one similar to the results obtained during electrical resistance measurements.

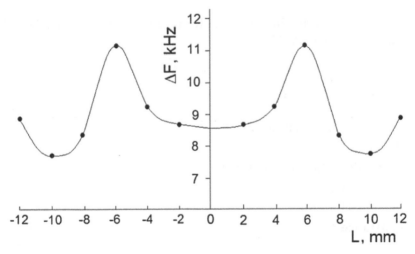

FIGURE 2.9   Magnetic susceptibility measurements of Co/C 50% polymer nanocomposite along the polymer composite film circular disk diameter.

When we move along the diameter of the disk, the frequency measurements indicate frequency change (frequency interval change of several kHz) associated with the change in the concentration of magnetic nanoparticles in the polymer composite.

Based on the obtained experimental results, it is possible to investigate the self-assembling processes in magnetic nanocomposite polymer films synthesized by the elaborated technology using the carbon magnetic nanopowders fabricated by a technology described in [4] under combined influence of magnetic field and heating.

2.3 CONCLUSION

The self-assembling processes were studied in the magnetic PNC synthe-sized based on the carbon nanoparticles doped with cobalt nanoclusters, synthesized by a unique CVD technology developed by the authors, under combined effect of magnetic field and heating. These processes took place with the diffusion of magnetic nanoparticles stimulated by the combined effect of outer steady magnetic field and heating. The obtained magnetic PNC have good electrical and adhesive properties and are promising for potential practical applications in magneto-electronics.

ACKNOWLEDGMENT

The work was supported by the Shota Rustaveli National Science Founda-tion and STCU grant # 6213.

KEYWORDS

- chemical vapor deposition
- diffusion
- magnetic carbon nanopowder
- polymer nanocomposites
- polyvinyl butyral
- resistance

REFERENCES

1. Jimenez-Contrares R., (2007). *Nanotechnology Research Developments*. Nova Science Publishers, Inc., New York.
2. King, V. R., (2007). *Nanotechnology Research Advances*. Nova Science Publishers, Inc., New York.
3. Grabowski, C. A., & Mukhopadhyay, A., (2014). *Macromolecules, 47*, 7238.
4. Gegechkori, T., Mamniashvili, G., Kutelia, E., et al., (2015). *J. Magn. Magn. Mater., 373*, 200.
5. Mette, C., Fischer, F., & Dilger, K., (2015). Proceedings of the institution of mechanical engineers. *Part L: Journal of Materials: Design and Applications, 229*(2), 166–172.
6. Aneli, J., Nadareishvili, L., & Shamanauri, L., (2016). *Journal of Electrical Engineering 4* (4), 196–202.

CHAPTER 3

Microbial Fuel Cell: A Green Microbial Approach for Production of Bioelectricity Using a Conducting Polymer

SWAPNALI A. PATIL, AKSHADA A. BAKLIWAL, and SWATI G. TALELE

Department of Pharmaceutics, Sandip Institute of Pharmaceutical Sciences, Nashik, Maharashtra, India, E-mail: swatitalele77@gmail.com (Swati G. Talele)

ABSTRACT

Microbial fuel cell (MFC) is one of the recently advanced sciences of using microbes for the generation of electricity. Microbial fuel cell undergoes an aerobiotic ferment of organic and inorganic matter to digest fuel which is obtained in energy form. This methodology is not so applicable due to expensiveness. MFC is an electrochemical science where chemical energy is converted to electrical energy using different microbes. In other words, MFC can also be titled as cellular respiration. The basic concept in is the reduction-oxidation respiration in the microbial culture which generates an energy in electrical form. The movement of electrons produces a potential difference which in turn helps to generate fuel. A path of hope for MFC conceptual is progressively studied to develop a new electrode in recent days. MFC is into a study to upgrade the concept with cost-effectiveness and helpful application of technique in new glance.

3.1 HISTORY

In 20th century, the idea of generating current from microbial cell came into highlight. M. C. Potter studied the concept in 1911 [1]. Potter produced electricity from the microbe *Saccharomyces cerevisiae*, but his work was not up to completion. Branet Cohen in the year 1931 made microbial half fuel cells which when was lined up in series was able to generate about 35 volts electric current [2].

A work by DelDuca et al. added hydrogen generated by the fermenting glucose by *Clostridium butyricum* as the reacting agent at the +ve end that is anode of a hydrogen along with air fuel cell. The cell worked out but it was not successfully carried out due to the owing to hydrogen instability by microbial production [3]. This complication was solved in 1976 by Suzuki et al. [4] who developed the MFC [5].

After in 1970 the concept of MFC was little, clear. The concept was put forward by Robin M. Allen and by H. Peter Bennetto. Population found the fuel cell as a possibility concept for the generating electric current in exploring countries. Bennetto's studies in 1980s, made a clear picture of operation of fuel cell came into focus.

The University of Queensland, Australia, in 2007 studied MFC as a collective effort with Foster's Brewing. The prototype, a 10 L design, transferred brewery wastewater into CO_2(carbon dioxide), pure water and electric current. The group planned to develop a pilot batch for presenting in an international conference on bioenergy [6].

3.2   INTRODUCTION

Due to growing population there is increase in necessitate of energy. So the new concept of deriving energy from bacterial has widely coming up. Microbial fuel cells (MFCs) is a process in which bacterium is selected as the catalyzer that does oxidation of organic and inorganic material and produce current flow [7–11]. The resources of energy are fuel fossil, nuclear energy and recycling resources depend on their sources and ability to exist. The nonrenewable resources are fossil fuel and nuclear energy. Unearthing and the carbon dioxide gas present under the earth and the fossil fuel are the reason for the global warming and environmental pollution [12]. For solving the problem of energy deficiency and problem associated with fossil fuel all countries of the world are trying to find a solution for recycling of energy sources. So MFC is one of the renewed techniques for production of electric energy by using metals as a catalyzer. MFC technique is one of the advance tech above the different energy-producing techniques to over the harmful oxides with greater effect. But the big drawback of this advanced technique is that the cost of application is too high [13]. An MFC is a process in which chemical energy is transformed formed by the oxidizing of organic and inorganic compound to produce adenosine triphosphate (ATP) by reacting with electrons to produce electric energy [14]. The common MFC consist of (+ve) Anode and (–ve) cathode part which is separated by a membrane. The microbes settle on anode and metabolism of organic compound such as glucose takes place. That acts as electron donor. The organic compound the electron and proton are produced by metabolism of organic compound. On the anode surface, the electron gets transferred.

From anode, the movement of electron to cathode takes place through the electrical circuit, while the protons move through the electrolyte and then reach the cationic membrane. To increase the speed of oxygen reduction on the cathode, Platinum is used as a catalyzer. Platinum electrodes are expensive the replacement of platinum electrode with other electrode does give proper oxygen reduction at cathode electrode [15, 16].

3.2.1   MFC DESIGNS

Different types of MFC are available. Most commonly use and less expensive is the two-chamber MFC developed on H pattern which has two bottles linked to each other by a tube which acts as a separator which is cation exchange membrane called Nafion, Ultex or plain salt bridge [17, 18]. The cation exchange membrane is also named as proton exchange membrane. In the H pattern, the proton exchange membrane is fixed in between the tubes of the connecting bottles. The two membranes can be kept together to get a large surface area. U shape tube heated and bent are used to join the bottles. The bottles are filled with agar and salt. Which acts as a cation exchange membrane. This method is less expensive. The salt bridge gives less amount of power because of high internal resistance. H shape pattern is used for initial research work for studying the generation of energy with different material or different microbial compounds. The surface area affects the amount of energy generated. H-shape type is the common parameter, such as studying power generation using recent available materials, of microbial communities that come up due to the degradation of certain compounds, but they commonly generate density of low power.

The power created relies upon the surface territory of cathode contrasted with that of anode [18] and the outside of the film [19]. The power thickness (P) produced by these systems is for the most part reliant on high inner opposition and cathode based misfortunes. When contrasting force produced by this sort, it makes to look at them based on similarly measured anodes, cathodes, and layers. By ferricyanide as an electron, acceptor at cathode chamber causes increment in the power thickness because of the great electron availability acceptor at high focuses. Due to Ferricyanide, causes increment in power by 1.5 to 1.8 occasions to a Pt-impetus cathode and broke down oxygen (H-structure reactor with a Nafion CEM). The most noteworthy thickness of intensity so far has detailed for MFC frameworks has low inside obstruction framework with ferricyanide at the cathode. While ferricyanide is a best catholyte as far as execution, it ought to be artificially recovered and its motivation of utilization isn't worth by and by. In this manner, the utilization of ferricyanide is restricted for central lab studies. It isn't important to put the cathode in water or in a different chamber while utilizing oxygen at the cathode. The cathode is in direct contact with air either in the nearness or nonappearance of a film [21]. In one sort of framework kaolin dirt, based separator and graphite cathode are joined to shape a consolidated separator-cathode structure [31]. High power densities have been gotten utilizing oxygen as the electron acceptor when watery cathodes are changed with air-cathodes. In the most straightforward framework type, the anode and cathode are put on either side of a cylinder, with the anode shut with a level plate and the cathode presented to air on one side, and water on the opposite side. At the point when a film is applied in this air-cathode framework, it conveys as fundamentally to repel water from spilling through the cathode, however, it additionally lessens oxygen dispersion at the anode chamber. The utilization of oxygen by microbes in the anode chamber can cause in a lower Coulombic productivity (which can be characterized as the portion of electrons secured as present versus the greatest conceivable recuperation) [21]. Hydrostatic weight on the cathode makes the break of water; however, that can be decreased by coatings, for example, polytetrafluoroethylene (PTFE), to the external side of the cathode which permits oxygen dispersion yet stops enormous loss of water [22]. Numerous minor departures from these structures have created in to attempt to improve control thickness or produce ceaseless course through the anode chamber (in change to the above frameworks which were altogether examined in group mode). Systems have been created with an external barrel-shaped reactor with a concentric inward cylinder that is at the cathode [23, 24], and with an internal round and hollow reactor (anode comprising of granular media) with the cathode outwardly [25]. Another change in plan the framework an up-flow fixed-bed biofilm reactor, with the liquid streaming persistently through permeable anodes toward a layer isolating the anode from the cathode chamber [26]. Frameworks have been created to look like hydrogen energy components, where a CEM is sandwiched between the anode and cathode. To expand the framework voltage,

MFCs can be contemplated with the frameworks structured in the arrangement of level plates or connected together in arrangement structure.

3.3   MANUFACTURING AND COST

The majority of the above have been exhibited from a fundamental science lab-based point of view, favorable with prior innovation arrange improvement, under controlled conditions. In any case, the best challenge for any innovation moving into this present reality is its reasonableness for assembling, which thus, drives economies of scale. The equivalent applies to the MFC improvement and this has been a piece of the test in the innovation taking off and becoming financially available. Most of the center parts and segments can be talked and along these lines costly, even at proto kind level, and there is a logical test in recognizing choices that would (i)perform similarly well and for delayed periods however, above all (ii)be economical and broadly accessible. One of the roads scientists have investigated is the elective ease materials including ceramic.

3.4   APPLICATION

One of the main applications could be the advancement of pilot-scale reactors at mechanical areas where a high caliber and solid influent are accessible. Nourishment handling wastewaters and digester effluents are great competitors. To look at the potential for power age at such a site, consider a nourishment preparing plant creating 7500 kg/d of waste organics in an emanating. This speaks to a potential for 950 kW of intensity, or 330 kW expecting 30% productivity. At an accomplished intensity of 1 kW/m^3, a reactor of 350 m^3 is required, which would generally cost 2.6 M Euros, at current costs. The created vitality, determined based on 0.1 Euros per kWh, is worth about 0.3 M Euros every year, giving ten-year restitution without different contemplations of vitality misfortunes or additions contrasted with other (high-impact) advancements. The reduction in production will reduce the payback time. In the long haul progressively weaken substrates, for example, household sewage, could be treated with MFCs, diminishing society's have to put generous measures of vitality in their treatment. A changed exhibit of elective applications could likewise rise, running from biosensor improvement and continued vitality age from the ocean bottom, to biobatteries working on different biodegradable powers. While full-scale, profoundly successful

MFCs are not yet inside our grip, the innovation holds impressive guarantee, and real obstacles will without a doubt be overwhelmed by designers and researchers. The increasing weight on our condition, and the call for sustainable power sources will further invigorate the advancement of this innovation, driving soon we would like to its effective usage.

3.4.1   GENERATION OF POWER

MFCs are alluring for power generation applications that require just low control, however where supplanting batteries might be unrealistic, for example, remote sensor networks [27–29]. Wireless sensors, controlled by microbial energy units can then for instance be utilized for remote checking (conservation) [30].

Basically, any natural material could be utilized to sustain the energy unit, including coupling cells to wastewater treatment plants. Compound procedure wastewater [31, 32] and engineered wastewater [33, 34] have been utilized to create bioelectricity in double and single-chamber mediators MFCs (uncoated graphite cathodes).

Higher power creation was seen with a biofilm-secured graphite anode [35, 36]. Fuel cell discharges are well under administrative limits [35]. MFCs use vitality more effectively than standard interior ignition motors, which are restricted by the Carnot Cycle. In principle, a MFC is fit for vitality proficiency a long way past 50% [36]. Rozendal acquired vitality change to hydrogen multiple times that of traditional hydrogen generation innovations.

MFCs can likewise work at a little scale. Anodes sometimes need just be 7 μm thick by 2 cm long [37] to such an extent that a MFC can supplant a battery. It gives an inexhaustible type of vitality and shouldn't be revived.

MFCs work well in gentle conditions, 20 °C to 40 °C and furthermore at pH of around 7 [38]. They do not require for long term medicinal applications, for example, in pacemakers. Power stations can be founded on amphibian plants, for example, green growth. Whenever sited neighboring a current power framework, the MFC framework can share its power lines.

3.4.2   EDUCATION

Soil-based microbial energy components fill in as instructive devices, as they include numerous logical orders (microbiology, geochemistry, electrical building, and so forth.) and can be made utilizing normally accessible materials, for example, soils, and things from the freeze. Packs for home science tasks and study halls are available. One case of microbial power devices being utilized in the study hall is in the IBET (Integrated Biology, English, and Technology) educational plan for Thomas Jefferson High School for Science and Technology. A few instructive recordings and articles are additionally accessible on the International Society for Microbial Electrochemistry and Technology (ISMET Society).

3.4.3 BIOSENSOR

The flow produced from a microbial power module is straightforwardly relative to the vitality substance of wastewater utilized as the fuel. MFCs can gauge the solute centralization of wastewater (i.e., as a biosensor). Wastewater is ordinarily evaluated for its biochemical oxygen demand (BOD) values. MFC BOD sensors underestimate BOD values in the presence of these electron acceptors normally actuated slop gathered from wastewater plants. A MFC-type BOD sensor can give constant BOD esteems. Oxygen and nitrate are favored electron acceptors over the terminal, lessening current age from a MFC. MFC BOD sensors think little of BOD esteem within the sight of these electron acceptors. This can be kept away from by hindering oxygen-consuming and nitrate breath in the MFC utilizing terminal oxidase inhibitors, for example, cyanide, and azide. Such BOD sensors are monetarily accessible.

The United States Navy is thinking about microbial power devices for ecological sensors. The utilization of microbial energy components to control natural sensors would almost certainly give capacity to longer periods and empower the accumulation and recovery of undersea information without a wired framework. The vitality made by these energy units is sufficient to support the sensors after an underlying startup time [34]. Due to undersea conditions (high salt focuses, fluctuating temperatures, and constrained supplement supply), the Navy may send MFCs with a blend of salt-tolerant microorganisms. A blend would consider a progressively complete usage of accessible supplements. *Shewanella oneidensis* is their essential competitor, however, may incorporate other warmth and cold-tolerant *Shewanella* spp. [35].

A first self-controlled and independent BOD/COD biosensor has been created and permits to distinguish natural contaminants in freshwater. The senor depends just on power delivered by MFCs and works constantly without upkeep. The biosensor turns on the caution to educate about sullying level: the expanded recurrence of the sign advises about higher defilement level, while low recurrence cautions about low tainting level [36].

3.4.4 BIORECOVERY

In 2010, Heijne et al. [37] Developed an instrument which was capable of generating electricity and reducing Cu (II) (ion) to copper metal. Microbial electrolysis cells have been constructed to produce hydrogen [38].

3.4.5 WASTEWATER TREATMENT

MFCs are being applicable in water treatment to harvest energy utilizing anaerobic digestion. The system can also decrease pathogens. However, it needs temperatures above of 30°C and requires an extra energy in order to convert biogas to electricity. Spiral spacers can be utilized to expand power generation by making a helical stream in the MFC. Scaling MFCs is an incredible test because of the power yield difficulties of a bigger surface region [39].

3.5 MATERIALS OF CONSTRUCTION

Anode–anodic materials must be conductive, biocompatible, and misleadingly stable in the reactor plan. Metal anodes comprising of noncorrosive tempered steelwork can be used [40], yet copper isn't valuable because of the harmfulness of even follow copper particles to microbes. The most adaptable anode material is carbon, accessible as smaller graphite plates, bars, or granules, as sinewy material (felt, fabric, paper, strands, froth), and as lustrous carbon. There are various carbon providers around the world, for instance, E_TEK and Electro combination Co. Inc. (USA), GEE Graphite Limited, Dewsbury (UK), Morgan, Grimbergen (Belgium), and Alfa-Aesar (Germany). The least complex materials for anode terminals are graphite plates or bars as they are generally economical, simple to deal with, and have a characterized surface zone. Graphite felt cathodes have large surface area [41] which can have high surface regions (0.47 m^2g^{-1}, GF series, GEE Graphite limited, Dewsbury, UK). In any case, not all the showed surface zone will fundamentally be accessible to microscopic organisms. Carbon fiber, paper, froth, and fabric (Toray) have been widely utilized as terminals.

It has been demonstrated that present increments with by and large inside surface region in the request carbon felt > carbon foam > graphite [42]. Generously higher surface region are accomplished either by utilizing a smaller material like reticulated vitreous carbon (RVC; ERG Materials and Aerospace Corp., Oakland, CA) [43] which is accessible with various pore sizes, or by utilizing layers of stuffed carbon granules (Le Carbone, Grimbergen, Belgium) or dots [48]. In the two cases keeping up high porosity is essential to avert stopping up. The long time impact of biofilm development or particles in the stream on any of the above surfaces has not been enough inspected. To expand the anode performance, diverse compound and physical systems have been pursued. Park et al. consolidated Mn (IV) and Fe(III) and utilized covalently connected unbiased red to intercede the electron move to the anode. Electrocatalytic materials, for example, polyanilins/Pt composites have additionally been appeared to improve the present age through helping the immediate oxidation of microbial metabolites.

Coordinating the watercourse through the anode material can be utilized to expand control. Cheng et al. found that stream coordinated through carbon fabric toward the anode, and diminishing terminal dividing from 2 to 1 cm, expanded power densities (standardized to the cathode anticipated surface territory) from 811 to 1540 mW/m^2 in an air-cathode MFC. The expansion was believed to be because of limited oxygen dissemination into the anode chamber, despite the fact that the advective stream could have assisted with proton transport toward the cathode also. Expanded power densities have been accomplished utilizing RVC in an up-flow UASB type MFC or in a granular anode reactor with ferricyanide cathodes. Move through an anode has likewise been utilized in reactors utilizing exogenous go-betweens cathode. Because of its great execution, ferricyanide (K3[Fe(CN)6]) is exceptionally well known as a test electron acceptor in microbial power modules. The best favorable position of ferricyanide is low over potential utilizing a plain carbon cathode, bringing about a cathode working potential near its open circuit potential. The best detriment, be that as it may, is the inadequate reoxidation by oxygen, which requires the catholyte to be consistently supplanted. Likewise, the long haul execution of the framework can be influenced by dispersion of ferricyanide over the CEM and into the anode chamber. Oxygen is the most reasonable electron acceptor for a MFC because of its high oxidation potential, accessibility, minimal effort (it is free), maintainability, and the absence of a compound waste item (water is shaped as the main end product). The decision of the cathode material incredibly influences execution, and is differed dependent on application. For residue power modules, plain graphite plate terminals drenched in the seawater over the silt have been utilized. Due to its extraordinary execution, ferricyanide (K3[Fe(CN)6]) is especially outstanding as a test electron acceptor in microbial power modules. The best good position of ferricyanide is the low over potential using a plain carbon cathode, achieving a cathode working potential close to its open circuit potential. The best impairment, in any case, is the insufficient reoxidation by oxygen, which requires the catholyte to be reliably displaced. Moreover, the whole deal execution of the system can be impacted by scattering of ferricyanide over the CEM and into the anode chamber. Oxygen is the most sensible electron acceptor for a MFC due to its high oxidation potential, openness, insignificant exertion (it is free), practicality, and the nonattendance of a compound waste thing (water is formed as the primary end product). The choice of the cathode material amazingly impacts execution, and is varied reliant on application. For buildup control modules, plain graphite plate terminals doused in the seawater over the residue have been used [51].

3.6   MANUFACTURING AND COST

The perspective, aim summary of the above study has being derived from past technology, research under unmaintained conditions. The wide scope which was the great challenge of the technique was to move into the present world and develop proper proved technique to carry with the economy of the world. Similarly, Microbial fuel cell is one of the challenging jobs in development and technology to make available in the commercial available in market. The development is a challenge to make the product available that to with cost reduction and widely and easily available. One of the option to explore and make the product develop at low cost is the material used for electrodes should be prepared from ceramic, cardboard [55, 56] or derivatives of plant. Geand Heinlong reported a long term pilot study in wastewater treatment plant which reduced the material cost by 60% of Microbial fuel cell because of cation exchanger membrane and so the expenses of the separator was reduced by using ceramic separators as a valid option for this technique. One of the important points of the parts availability like electrodes, membrane, wiring to compile a large scale. The technique developed in laboratory scale need to be expanded in pilot study to make the availability of the technique assessable to the real world under different environmental condition such as temperature change, humidity, batch, and continuous flow pattern. The institutional level knowledge would be a boon in the microbial fuel cell in real-time. Microbial fuel cell reactors and grade can produce electricity spontaneously, potentially, effectively with cost reduction, energy consumption, and treatment cycle. The energy from the vitality investigation perspective, 14 Wh achievedatthe last Pee Power field preliminary at Glastonbury-2017 would be the equivalent to 01.23 British pence of vitality putting something aside for each kWh. This depends on crude power information delivered during the field preliminary where the MFC stack was worked on perfect human urine and doesn't consider the sparing that would be picked up for each liter of wastewater treated. Further comprehension of particle transport selectivity and financial layer arrangement techniques are imperative to en-capable more extensive work of particle trade films in technical forms for manageable improvement. Further progress is expected to give field hardware that is increasingly powerful and solid additional time just as the development of novel vitality stockpiling and vitality collecting strategies. The stored energy is applied as external capacitor is new path in this field. The study complied on the present data is an attempt of new concept in the market. Surface modification of the anode materials also contribute to the boosted up performance of the MFCs. A carbon cloth was modified by coating with biologically reduced Palladium nanoparticles to develop a functional anode. With this surface-modified carbon cloth, MFC has shown improved power density and columbic efficiencies by 14% and 31% respectively. The better execution could be because of the diminished charge move obstruction by the biogenic palladium. Mink et al. have built up an anode of vertically developed multiwall carbon nanotubes (CNTs) with nickel silicide. This manufactured terminal delivered a current thickness of 197 mA/m^2 and a power thickness of 392 mW/m^3 in a micronized MFC of 1.25 µL limit. The outcomes were nearly higher than that accomplished with a plain carbon material. It was asserted that the upgraded presentation was credited to expanded surface to volume proportion for microbial bond, electron move, and low opposition by the nickel silicide reaching material. It has been reasoned that since the procedure doesn't include any costly materials to manufacture MFC, it can significantly diminishes the related expenses. Carbon paper modified with multiwall CNTs and polyethyleneimine through layer by layer assembled technique was used as anode in an MFC. The modification has brought about reduction in interfacial charge transfer through providing three-dimensional network structures for bacterial adhesion. This modified anode was capable of producing 20% higher power densities than an unmodified carbon cloth electrode. Stainless steel cross-sections covered with warmth treated and untreated goethites acquired from mining mud were utilized as anode materials in MFC applications. Higher powers and columbic efficiencies were recorded with goethite catalyst based electrode than conventional stainless steel mesh electrode. The improved performance was due to reduced mass transfer losses and accelerated electron transfer between biocatalysts and electrode. Goethite being cheaper can provide the economically feasible electrode materials with acceptable power generations in MFC. The use of carbon cloth or graphite granules as anodes in lab-scale MFC even though showed better performances, still combating the challenges of scale-up. Carbon cloth being expensive cannot be used in large scale systems. In the same way, well-establishedlab-scale configurations of carbon cloth would be difficult to use in large scales. Similarly, graphite granules are vulnerable to clogging due to low porosities. Materials to be used in biofilm-based bioreactors (MFCs) for wastewater treatment must have high structural strength and should be free from clogging since, they are to be in constant contact with the electron transfer. In addition, the material supports should be intact with the biofilm under any operational condition from open-flow to saturated flow systems. Hence, the use of such support materials is impractical. To overcome the clogging issues, Logan et al. have configured a new anode with high surface area graphite fiber brush bounded to a central core collecting material. The design has led to the development of a successful high conducting and high specific surface area anode which can be used for scale-up MFC applications. 4.2 Cathode materials: In MFC, the cathode chamber is considered as a sink of electrons, in which oxygen is reduced to water. In cathode chamber, reduction of oxygen occurs at air, liquid, and solid three-phase interface. A typical MFC cathode is composed of an electrode support, catalyst, and air diffusion layer. Electrode materials used for anode can also be used for cathode but a potential cathode should have the properties of high electric conductivity, high mechanical strength, and effective catalytic nature. Most commonly, MFCs are being operated under neutral PH and mild temperatures. At these conditions, the rate of oxygen reduction is very low, leading to increase over potentials and thereby limiting the performance of an MFC. For robust cathodic reactions in MFCs, the cathodic carbon-based support materials must be amended with additional catalysts.

Platinum is most regularly utilized cathode impetus because of its high productivity towards oxygen decrease. The utilization of costly metal impetuses as cathode materials in MFCs is obliging the viable uses of MFC innovation. Notwithstanding significant expense, Platinum impetuses are progressively at risk of fouling when utilized with low-quality water. A lot of research endeavors have been put to diminish the expense of cathodic impetuses by finding less expensive options in contrast to Platinum without relinquishing the exhibition. This audit talks about the financially savvy cathode materials and their impact on MFC execution. In one investigation, the carbon material was covered on one side with a blend of carbon powder and poly-tetra-fluoro-ethylene base layer pursued by a few coatings of poly-tetra-fluoro-ethylene as dispersion layer. The water confronting side of the carbon fabric was covered with platinum impetus. In this way, manufactured carbon material was utilized as cathode in a solitary chamber MFC. The MFC has indicated upgraded control densities (42%) and columbic efficiencies (200%) when contrasted with carbon material with base layer alone. It has been accounted for that four coatings of poly-tetra-fluoro-ethylene delivered greatest power thickness as it can diminishes the water misfortunes through cathode. Cathode impetuses made of metal porphyrins and phthalocyanines bolstered on Ketjen black carbon were read for their synergist action towards oxygen decrease in MFCs. Iron phthalocyanine based cathode has indicated more noteworthy oxygen decrease rates at unbiased PH than Platinum impetus in impartial PH. On looking at the impact of carbon substrate, Ketjen black carbon has revealed preferable action over Vulcan XC carbon because of higher surface region of Ketjen black carbon. Most extreme power thickness of 634 mW/m^2 was accomplished with Iron phthalocyanine-Ketjen black carbon at unbiased pH, which is higher than that of costly Platinum impetus (593 mW/m^2) at comparable conditions. It has been concluded from the examinations that the progress metal-based macrocyclicimpetuses are less expensive and can be effectively utilized in enormous scale MFCs for commercialization of the innovation. In a way to deal with create financially savvy interchange to Platinum impetus, Zhang et al. have created initiated carbon air cathode. The cathode was created by virus squeezing actuated carbon with poly-tetra-fluoro-ethylene around a nickel work current authority. The firm has announced the greatest power thickness of 1220 mW/m^2 with the subsequently manufactured cathode, which is relatively a lot higher than that acquired with platinum impetus (1060 mW/m^2). From the outcomes, it was finished up by the gathering that initiated carbon-metal work authority determined cathodes can be utilized as productive and monetarily possible air cathodes in MFCs. Another firm has revealed that initiated carbon-poly-tetra-fluoro-ethylene air cathodes have demonstrated upgrades in power densities with the expansion in actuated carbon stacking. This could be because of the decreases in Warburg impedance, contact obstruction, and accuse move opposition of the actuated carbon load. Regardless of cathodes debasement over 1.5–5 months of utilization, still these cathodes displayed control densities similar to platinum-based cathodes. In an alternate report business actuated carbons from various source materials were alkali gas treated to improve the exhibition as oxygen lessening impetuses in MFCs at unbiased PH. Smelling salts gas treated initiated carbon impetuses have come about preferred exhibitions over untreated enacted carbon on account of the decrease in oxygen substance and improvement in nitrogen bunches on the outside of the impetus material. To upgrade the cathode reactant action of initiated carbon, Xia et al. have built up an actuated carbon-Iron ethylene di-amine tetra-acetic corrosive cathode with tempered steel work current authority. In this way created cathode has displayed altogether improved power thickness (1580 ± 80 mW/ m^2), which is 10% more noteworthy than plain actuated carbon cathode, and furthermore, the power thickness is tantamount to platinum impetus based cathode. It was accounted for that the anode is tougher than platinum cathodes. The improved exhibition of the carbon-Iron ethylene di-amine tetra-acetic corrosive cathode was seen as because of the dynamic pyridine, quaternary nitrogen, and iron gatherings. The investigation has announced that the reactant movement of financially feasible enacted carbon cathodes will be improved by pyrolyzing actuated carbon with Iron ethylene di-amine tetra-acetic corrosive. Afterward, Zhang et al. have mixed enacted carbon with carbon dark to lift up the presentation and life span of initiated carbon cathodes. It was accounted for that the most extreme power thickness was diminished by 7% for the carbon dark based cathode while, 61% decrease was accounted for with platinum impetus following five months of ceaseless activity.

Zuo and Logan have examined the power densities of carbon cloth by changing the configurations (tube/flat). Initially, use of a single carbon cloth tube resulted in lower power density than a flat carbon cloth electrode. It was noticed that with the increase in number of concentric tubes, the power density increased due to increase in the cathode surface area. The firm has obtained a power density of 83 mW/m^3 with two carbon cloth tubes as cathode. It was also reported that on wrapping the cathode around anode like a tubular MFC, power density increased to 128 mW/m^3. Finally, from the analysis of the results, it has been concluded that geometry of cathode either tube or flat will not affect the power density. The power density is a dependent factor of cathode surface area rather than geometry. Generally, major costs of MFC cathodes are ascribed by catalysts and catalyst binders. Packed bed air cathodes were developed with four carbon-based materials (granular activated carbon, granular semi-coke, carbon felt cubes, and granular graphite) and the fabrication was devoid of expensive binders and catalysts. Granular activated carbon-based packed bed air cathode has shown the highest power density of 676 ± 93 mW/m^2 and lowest was reported with carbon felt based cathode (60 ± 43 mW/m^2). It was observed that on increasing the amount of granular activated carbon and semicoke with an intention to increase the surface area, power generation was significantly dropped. Thicker layers of carbon materials limited the diffusion of oxygen, thereby reducing the oxygen reduction rate. From the reports, it was deduced that packed bed air cathodes of granular activated carbon and granular semi coke can be used as potential cost-effective alternates to platinum air cathodes. Mesoporous nitrogen-rich carbon developed at different temperatures was used as cathode material to address the issue associated with an expensive platinum catalyst. Mesoporous nitrogen-rich carbon cathode reported 14% less power production than platinum cathode but showed only a 7% reduction after a month operation when it was 11% for platinum cathode. Zhang et al. [49] have experimented with nitrogen-dopedionothermal carbon aerogel as air cathode to improve the rate of oxygen reduction. This aerogel based electrode resulted in 1.7 times higher power density than usual platinum electrode as well as most of the oxygen reducing catalyst air cathodes due to highly porous nature, high surface area and large pore volume. The firm has reported that the proposed aerogel electrode could be an efficient cost-effective air cathode catalyst for MFC applications. In recent years, tremendous studies have been carried out towards the development of design and electrode materials to boost up the performance of MFCs. However, still the technology is facing the challenges for commercialization. In sum up, stringent research is needed to further develop the hardware.

3.7 OUTLOOK

MFC plans need enhancements before an attractive item will be conceivable. Both the issues recognized above and the scale-up of the procedure stay basic issues. The greater part of the structures investigated here can't be scaled to the level required for an enormous wastewater treatment plant which requires many cubic meters of reactor volume. Either the characteristic change pace of MFCs should be expanded, or the plan should be improved so a savvy, huge scale framework can be created. Plans that can most effectively be made in stacks, to create expanded voltages, will be helpful as the voltage for a solitary cell is low. The achievement of explicit MFC applications in wastewater treatment will rely upon the fixation and biodegradability of the natural issue in the influent, the wastewater temperature, and the nonattendance of lethal synthetics. Materials costs will be a huge factor in the all-out reactor costs. For the most part, anodic materials generally utilized in MFC reactors, for example, graphite froths, reticulated vitreous carbon, graphite, and others, are very costly. Rearranged cathodes, for example, carbon strands, may reduce these anode costs. The utilization of costly impetuses for the cathode should likewise be stayed away from. Another essential perspective is the evacuation of non-carbon-based substrates from the waste streams: nitrogen, sulfur, and phosphorus-containing mixes frequently can't be released into nature at influent focuses. So also, even particulate natural mixes should be evacuated and changed over to effectively biodegradable mixes, as a feature of a successful wastewater treatment activity.

Soil-based microbial energy units fill in as instructive apparatuses, as they incorporate various logical orders (microbiology, geochemistry, electrical building, and so on.) and can be made utilizing usually accessible materials, for example, soils, and things from the cooler. Units for home science tasks and study halls are accessible. One case of microbial power devices being utilized in the homeroom is in the IBET educational program for Thomas Jefferson High School for Science and Technology. A few instructive recordings and articles are likewise accessible on the International Society for Microbial Electrochemistry and Technology (ISMET Society).

3.8 CONCLUSION

MFC is a cutting edge innovation for creation of power from digestion of microorganisms. In this survey, we have managed significant squanders and xenobiotic, for example, hexavalent chromium, agro wastes, nitrates, and azo colors. Some of them, for example, hexavalent chromium, and azo colors are exceptionally lethal to the biological system and cause demise of living beings. In MFC, they are utilized for power generation and furthermore they are changed into less harmful metabolites, which exhibits it's another potential use in waste administration and contamination control. Till now, countless microorganisms and a waste assortment of substrates (counting waste and xenobiotic) have been utilized to deliver power. Be that as it may, a noteworthy disadvantage of this innovation is that the power yield is extremely low and scaling up prompts a lessening in power yield. This is the primary motivation behind why this innovation has yet not been marketed. Thus, much more work is required with the goal that this innovation ends up effective, material, and broadly acknowledged.

KEYWORDS

- **adenosine triphosphate**
- **aerobiotic ferment**
- **anode**
- **biochemical oxygen demand**
- **cathode**
- **microbial fuel cells**

REFERENCES

1. Potter, M. C., (1911). Electrical effects accompanying the decomposition of organic compounds. *Proceedings of the Royal Society B: Biological Sciences, 84*(571), 260–276.
2. Cohen, B., (1931). The bacterial culture as an electrical half-cell. *Journal of Bacteriology, 21*, 18–19.
3. DelDuca, M. G., Friscoe, J. M., & Zurilla, R. W., (1963). Developments in industrial microbiology. *American Institute of Biological Sciences, 4*, 81–84.
4. Karube, I., Matasunga, T., Suzuki, S., & Tsuru, S., (1976). Continuous hydrogen production by immobilized whole cells of *Clostridium butyricum. Biochimica et Biophysica Acta, 24*(2), 338–343.
5. Karube, I., Matsunaga, T., Tsuru, S., & Suzuki, S., (1977). Biochemical cells utilizing immobilized cells of *Clostridium butyricum. Biotechnology and Bioengineering, 19*(11), 1727–1733.
6. Jurg Keller, & Rabaey, (2014). *Brewing a Sustainable Energy Solution.* The University of Queensland Australia.
7. Berk, R. S., & Canfield, J. H., (1964). Bioelectrochemical energy conversion. *Appl. Microbiol., 12*, 10–12.
8. Rao, J. R., Richter, G. J., Von, S. F., & Weidlich, E., (1976). The performance of glucose electrodes and the characteristics of different biofuel cell constructions. *Bioelectrochem. Bioenerg., 3*, 139–150.
9. Davis, J. B., & Yarbrough, H. F., (1962). Preliminary experiments on a microbial fuel cell. *Science, 137*, 615–616.
10. Cohen, B., (1931). The bacterial culture as an electrical half-cell. *J. Bacteriol., 21*, 18, 19.

11. Potter, M. C., (1911). Electrical effects accompanying the decomposition of organic compounds. *Proc. R. Soc. London Ser. B, 84,* 260–276. https://doi.org/10.1098/rspb.1911.0073

12. Rahimnejad, M., Adhami, A., Darvari, S., Zirepour, A., & Oh, S. E., (2015). Microbial fuel cell as new technology for bioelectricity generation: A review. *AEJ, 54,* 745–756.

13. Peighambardoust, S., Rowshanzamir, S., & Amjadi, M., (2010). Review of the proton exchange membranes for fuel cell application. *Int. J. Hydrog Energy, 35,* 9349–9384.

14. Torres, C. I., Brown, R. K., Parameswaran, P., Marcus, A. K., Wanger, G., Gorby, Y. A., & Rittmann, B. E., (2009). Selecting anode-respiring bacteria based on anode potential: Phylogenetic, electrochemical, and microscopic characterization. *Environ Sci. Technol., 43,* 9519–9524.

15. Sun, J., Bi, Z., Hou, B., Cao, Y., & Hu, Y., (2011a). Further treatment of decolorization liquid of azo dye coupled with increased power production using microbial fuel cell equipped with an aerobic biocathode. *Water Res., 45,* 283–291.

16. Sun, J., Hu, Y. Y., & Hou, B., (2011b). Electrochemical characterization of the bioanode during simultaneous azo dye decolorization and bioelectricity generation in an air-cathode single chambered microbial fuel cell. *Electrochim. Acta, 56*(19), 6874–6879.

17. Bond, D. R., Holmes, D. E., Tender, L. M., & Lovley, D. R., (2002). Electrode reducing micro organisms that harvest energy from marine sediments. *Science, 295,* 483–485.

18. Park, D. H., (1999). Utilization of electrically reduced neutral red by *Actinobacillus succino genes*: Physiological function of neutral red in membrane-driven fumarate reduction and energy conservation. *J. Bacteriol., 181,* 2403–2410.

19. Oh, S., Min, B., & Logan, B. E., (2004). Cathode performance as a factor in electricity generation in microbial fuel cells. *Environ. Sci. Technol., 38,* 4900–4904.

20. Oh, S., & Logan, B. E., (2006). Proton exchange membrane and electrode surface areas as factors that affect power generation in microbial fuel cells. *Appl. Microbiol. Biotechnol., 70,* 162–169.

21. Liu, H., & Logan, B. E., (2004). Electricity generation using an air-cathode single chamber microbial fuel cell in the presence and absence of a proton exchange membrane. *Environ. Sci. Technol., 38,* 4040–4046.

22. Cheng, S., Liu, H., & Logan, B. E., (2006). Increased performance of single chamber microbial fuel cells using an improved cathode structure. *Electrochem. Commun., 8,* 489–494.

23. Habermann, W., & Pommer, E. H., (1991). Biological fuel cells with sulfide storage capacity. *Appl. Microbiol. Biotechnol., 35,* 128–133.

24. Liu, H., Ramnarayanan, R., & Logan, B. E., (2004). Production of electricity during wastewater treatment using a single chamber microbial fuel cell. *Environ. Sci. Technol., 38,* 2281–2285.

25. Rabaey, K., Clauwaert, P., Aelterman, P., & Verstraete, W., (2005). Tubular microbial fuel cells for efficient electricity generation. *Environ. Sci. Technol., 39,* 8077–8082.

26. He, Z., Minteer, S. D., & Angenent, L. T., (2005). Electricity generation from artificial wastewater using an up flow microbial fuel cell. *Environ. Sci. Technol., 39,* 5262–5267.

27. Subhas, C. M., & Joe-Air, J., (2013). Application of microbial fuel cells to power sensor networks for ecological monitoring. *Wireless Sensor Networks and Ecological Monitoring: Smart Sensors, Measurement, and Instrumentation* (Vol. 3, pp. 151–178). Springer link.

28. Wang, V. B., Song-Lin, C., Cai, Z., Sivakumar, K., Zhang, Q., Kjelleberg, S., Cao, B., et al., (2014). A stable synergistic microbial consortium for simultaneous azo dye removal and bioelectricity generation. *Bioresource Technology, 155*, 71–76.

29. Wang, V. B., Song-Lin, C., Cao, B., Seviour, T., Nesatyy, V. J., Marsili, E., Kjelleberg, S., et al., (2013). Engineering PQS biosynthesis pathway for enhancement of bioelectricity production in Pseudomonas aeruginosa microbial fuel cells. *PLoS One, 8*(5), e63129.

30. Venkata, M. S., Mohanakrishna, G., Srikanth, S., & Sarma, P. N., (2008). Harnessing of bioelectricity in microbial fuel cell (MFC) employing aerated cathode through anaerobic treatment of chemical wastewater using selectively enriched hydrogen producing mixed consortia. *Fuel, 87*(12), 2667–2676.

31. Venkata, M. S., Mohanakrishna, G., Reddy, B. P., Saravanan, R., & Sarma, P. N., (2008). Bioelectricity generation from chemical wastewater treatment in mediator less (anode) microbial fuel cell (MFC) using selectively enriched hydrogen producing mixed culture under acidophilic microenvironment. *Biochemical Engineering Journal, 39*, 121–130.

32. Mohan, S. V., Veer, R. S., Srikanth, S., & Sarma, P. N., (2007). Bioelectricity production by mediator less microbial fuel cell under acidophilic condition using wastewater as substrate: *Influence of substrate loading rate. Current Science, 92*(12), 1720–1726. JSTOR24107621.

33. Venkata, M. S., Saravanan, R., Raghavulu, S. V., Mohanakrishna, G., & Sarma, P. N., (2008). Bioelectricity production from wastewater treatment in dual chambered microbial fuel cell (MFC) using selectively enriched mixed micro flora: Effect of catholyte. *Bioresource Technology, 99*(3), 596–603.

34. Venkata, M. S., Veer, R. S., & Sarma, P. N., (2008). Biochemical evaluation of bioelectricity production process from anaerobic wastewater treatment in a single chambered microbial fuel cell (MFC) employing glass wool membrane. *Biosensors and Bioelectronics, 23*(9), 1326–1332.

35. Venkata, M. S., Veer, R. S., & Sarma, P. N., (2008). Influence of anodic biofilm growth on bioelectricity production in single chambered mediator less microbial fuel cell using mixed anaerobic consortia. *Biosensors and Bioelectronics, 24*(1), 41–47.

36. Choi, Y., Jung, S., & Kim, S., (2000). *Development of Microbial Fuel Cells Using Proteus Vulgaris Bulletin of the Korean Chemical Society, 21*(1), 44–48.

37. Ting Chen, Scott Calabrese Barton, Gary Binyamin, Zhiqiang Gao, Yongchao Zhang, Hyug-Han Kim, & Adam Heller, (2001). A miniature biofuel cell. *J. Am. Chem. Soc., 123*(35), 8630–8631.

38. Bullen, R. A., Arnot, T. C., Lakeman, J. B., & Walsh, F. C., (2006). Biofuel cells and their development(PDF). *Biosensors and Bioelectronics, 21*(11), 2015–2045.

39. Tanisho, S., Kamiya, N., & Wakao, N., (1989). Microbial fuel cell using *Enterobacter aerogenes. Bioelectrochem. Bioenerg., 21*, 25–32.

40. Gil, G. C., Chang, I. S., Kim, B. H., Kim, M., Jang, J. K., Park, H. S., & Kim, H. J., (2003). Operational parameters affecting the performance of a mediator-less microbial fuel cell. *Biosens. Bioelectron., 18*, 327–334.

41. Chaudhuri, S. K., & Lovley, D. R., (2003). Electricity generation by direct oxidation of glucose in mediator less microbial fuel cells. *Nat. Biotechnol., 21*, 1229–1232.

42. Kim, N., Choi, Y., Jung, S., & Kim, S., (2000). Development of microbial fuel cell using *Proteus vulgaris. Bull. Korean Chem. Soc., 21*, 44–49.

43. Sell, D., Kramer, P., & Kreysa, G., (1989). Use of an oxygen gas diffusion cathode and a three-dimensional packed bed anode in a bioelectrochemical fuel cell. *Appl. Microbiol. Biotechnol., 31*, 211–213.

44. Lowy, D. A., Tender, L. M., Zeikus, J. G., Park, D. H., & Lovley, D. R., (2006). Harvesting energy from the marine sediment-water interface II-Kinetic activity of anode materials. *Biosens. Bioelectron., 21*, 2058–2063.

45. Niessen, J., Schroder, U., Rosenbaum, M., & Scholz, F., (2004). Fluorinated polyanilines as superior materials for electrocatalytic anodes in bacterial fuel cells. *Electrochem. Commun., 6*, 571–575.

46. Schroder, U., Niessen, J., & Scholz, F., (2003). A generation of microbial fuel cells with current outputs boosted by more than one order of magnitude. *Angew. Chem., Int. Ed., 42*, 2880–2883.

47. Cheng, S., Liu, H., & Logan, B. E., (2006). Increased power generation in a continuous flow MFC with advective flow through the porous anode and reduced electrode spacing. *Environ. Sci. Technol., 40*, 2426–2432.

48. Bergel, A., Feron, D., & Mollica, A., (2005). Catalysis of oxygen reduction in PEM fuel cell by seawater biofilm. *Electrochem. Commun., 7*, 900–904.

49. Cheng, S., Liu, H., & Logan, B. E., (2006). Power densities using different cathode catalysts (Pt and CoTMPP) and polymer binders (Nafion and PTFE) in single chamber microbial fuel cells. *Environ. Sci. Technol., 40*, 364–369.

50. Zhao, F., Harnisch, F., Schroder, U., Scholz, F., Bogdanoff, P., & Herrmann, I., (2005). Application of pyrolyzed iron(II) phthalocyanine and CoTMPP based oxygen reduction catalysts as cathode materials in microbial fuel cells. *Electrochem. Commun., 7*, 1405–1410.

51. Rabaey, K., & Verstraete, W., (2005). Microbial fuel cells: Novel biotechnology for energy generation. *Trends Biotechnol., 23*, 291–298.

52. Subhas, C. M., & Joe-Air, J., (2013). Application of microbial fuel cells to power sensor networks for ecological monitoring. *Wireless Sensor Networks and Ecological Monitoring: Smart Sensors, Measurement, and Instrumentation* (Vol. 3, pp. 151–178). Springer link.

53. Wang, V. B., Song-Lin, C., Cai, Z., Sivakumar, K., Zhang, Q., Kjelleberg, S., Cao, B., et al., (2014). A stable synergistic microbial consortium for simultaneous azo dye removal and bioelectricity generation. *Bioresource Technology, 155*, 71–76.

54. Wang, V. B., Song-Lin, C., Cao, B., Seviour, T., Nesatyy, V. J., Marsili, E., Kjelleberg, S., et al., (2013). Engineering PQS biosynthesis pathway for enhancement of bioelectricity production in *Pseudomonas aeruginosa* microbial fuel cells. *PLoS One, 8*(5), e63129.

55. Venkata, M. S., Mohanakrishna, G., Srikanth, S., & Sarma, P. N., (2008). Harnessing of bioelectricity in microbial fuel cell (MFC) employing aerated cathode through anaerobic treatment of chemical wastewater using selectively enriched hydrogen producing mixed consortia. *Fuel, 87*(12), 2667–2676.

56. Venkata, M. S., Mohanakrishna, G., Reddy, B. P., Saravanan, R., & Sarma, P. N., (2008). Bioelectricity generation from chemical wastewater treatment in mediator less (anode) microbial fuel cell (MFC) using selectively enriched hydrogen producing mixed culture under acidophilic microenvironment. *Biochemical Engineering Journal, 39*, 121–130.

CHAPTER 4

Sustainability of the Tire Industry: Through a Material Approach

SANJIT DAS,[1] HIRAK SATPATHI,[1] S. ROOPA,[2] and SAIKAT DAS GUPTA[1]

[1]Hari Shankar Singhania Elastomer and Tire Research Institute (HASETRI), Plot No 437, Hebbal Industrial Area, Mysore, Karnataka – 570016, India

[2]Department of Polymer Science and Technology, Sri Jayachamarajendra College of Engineering, JSS Science and Technology University, Mysore – 570006, India

ABSTRACT

The tire industry is basically a slow-changing industry, however, in the present era of fast change, a number of environmental and sustainability-related issues are impacting the tire industry to a large extent. Non-pneumatic tire, tires for electric vehicles, tires with sensors is talk of the day to meet the future requirements of the automobile industries. All these tires need to have fuel efficiency, safety at high speed, and low noise. Tire designers and material scientists are jointly involved in tire design, material design to achieve targeted tire performance properties. Material scientists are focusing their research mainly on sustainable materials as an alternate to present fossil-based material. In this chapter, authors will share different research work are being carried out on sustainable materials particularly for the tire industry. Reuse-Regenerate-Recycle is one of the research areas to reduce the consumption of virgin fossil fuel-based materials. Natural rubber (NR) is one of the commonly used rubbers in the tire industry. It is a naturally occurring material and helps the tire industry to achieve sustainability in the long run. However, the imbalance between the demand and supply of NR, worldwide scientists are doing research on Guayule, Dandelion to obtain a suitable alternative of NR. With the introduction of stringent requirements of tire labeling criteria by the European Union, materials scientists are bound to use alternative of carbon black reinforcing filler. One of the major alternatives of carbon black is silica which is also an eco-friendly sustainable material. Synthetic rubbers using plant-derived materials or biomass (agricultural waste) are came into the picture, Liquid Farnesene rubber is also another newly introduced polymer from bio-origin. Research efforts in developing microbes that can grow an isoprene monomer used to make a mimics NR is also undergoing. Fillers which are the next bulk consuming material after polymer are being explored from non-petroleum-based origin like silica, clay, and from bio-origin like rice husk ash, cornflower starch, bagasse; eggshell derived nano calcium carbonate. Coupling agents like Baker's yeast as an alternate of TESPT. Process aids from non-petroleum-based origin like soybean oil, orange oil, canola oil, sunflower oil, neem oil, etc. are also under the different stages of research for application in the tire. Apart from that, many resins from natural source are used to replace petroleum-based process oil and also to enhance tire performance. Cellulose-based synthetic fibers are of profound interest to replace synthetic fiber in tire reinforcement. Researchers are working to get anti-oxidant from flowers, fruits peels, and food wastes to replace petro-based antioxidants. Tea polyphenols as anti-oxidant has already found potential for the rubber compound. Investigations are also being done for a single compounding ingredient which would possess several functions and thus the term surfactant-accelerator-processing aid (SAPA) or termed as multi-functional additives (MFA).

4.1   INTRODUCTION

Sustainability and sustainable development are the key focus area of technology, research, and development of any industry. Sustainability implies meeting the present needs without compromising the freedom of future generations need. The main goal of sustainability is to comply with three requirements such as environment, economy, and society [1] (see Figure 4.1). New regulations from different government bodies, environmental bodies forces all the industrial sectors to adhere to the norms to make the environment more sustainable by reducing carbon footprints. Sustainability seeks an improved quality of life and embraces equality for all [2]. Sustainable development aims at assuring the ongoing productivity of exploitable natural resources and conserving all species of fauna and flora [3], Brundtland Commissions (United Nations) 1987.

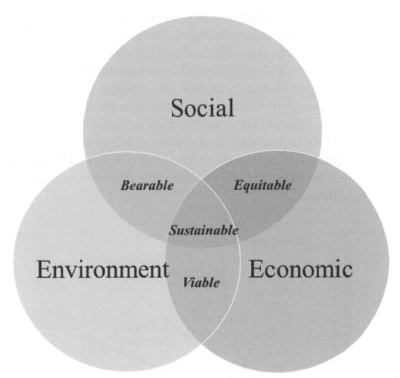

FIGURE 4.1   Graphical representation of sustainability (Venn diagram of sustainable development).

According to Philip Sutton [4], environmental sustainability can be achieved by reducing usage of physical resources and using more and more renewable rather than depletable resources. To achieve this, redesign of product and process is necessary to avoid toxic material generation. Sustainability goals and targets are universal. Reaching the goals is spread over all fonts-governments, business, civil society, and people everywhere to play the role. Currently, the way goods and services are produced; it is unsustainable and contributes significantly towards environmental problems. A new focus on sustainable developments needs new approach. Figure 4.2 illustrates the transition from business as usual to sustainable development [5].

Private sectors have to understand that sustainability may be breaking even or accommodate a loss in the short-term but if done correctly, then it can be profitable in long-term. Collaboration across sectors is inevitable to handle big issues. Breaking of silos in between sectors and collaboration between different stakeholders-suppliers, customers, partners in value chain, universities, and research institutes is one of the ways to reach the goal of sustainability. Sustainability is growing demand for tire producers. Eventually chemical industry producing raw materials has the big challenge as well. In rubber industry, many types of gases, fumes, and vapors come out of chemicals during use and storage and also during high temperature curing. These toxic materials ultimately pollute environment. Almost all rubber manufacturer discharge oil mixed water, chemical contaminated water from testing labs. These effluent waters pollute drinking water resources, land of irrigation [6]. REACH regulation and restriction brings some measure and put band on some toxic chemicals which is being followed by all reputed rubber manufacturers and ultimately some respite from environmental pollution level is achieved.

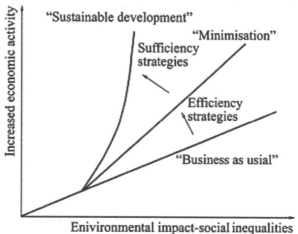

FIGURE 4.2   Shifting development path towards sustainable development.

Source: Reprinted with permission from Ref. [5]. © 2010 Elsevier.

According to new report "Plastic and Climate" published in 2019, plastic will contribute greenhouse gases equivalent of 850 million tons of carbon dioxide (CO_2). In current trend, annual emissions will reach 1.34 billion tons by 2030. By 2050, plastic could emit 56 billion tons of greenhouse gas emissions, as much as 14% of the earth's remaining carbon budget. Published research in 2019, showing that insects are destroyed by human activities like habitat destruction, pesticide poisoning, invasive species, and climate change (CC). Unless it cannot be stopped, it will result in ecological imbalance on earth in the next 50 years. Presently we are following linear material economy. Our material using fashion is characterized by the sequence take-make-use-scrap. To achieve sustainable goal, present days novel idea of circular economy emphasizes on 3R, i.e., repair-reuse-recycle. Raw Material availability and cost is going to affect business profoundly. Population growth and economic development provide spurs and opportunities for business, especially as it relates to massive developing countries China, India, and Brazil. Sustainable materials are those whose use brings benefits, unlike conventional materials. It must be naturally abundant, an ease of extraction with minimum energy spent and easy to recycle. Two general classes may be categorized. First, one is those materials which are mainly from plant origin like wood, natural fibers and polymers and the second one is mostly recycled materials, generally produced from waste products and raw materials. Automobiles uses report says by 2017 approx. 78.6 million of passenger cars are in use worldwide [7]. Considering approx. 6.5 billion people, around 87% of world population is to receive benefits of automobile [1]. The number of vehicles (excluding motorcycle) in use around the world is expected to reach 1.2 billion by 2020 and 1.6 billion in 2030 due to increase demand in developing countries [8–10]. Although modern automobile elastomers component is 8% by weight, but it is one of the key materials from engineering and design viewpoint [11]. With increase of vehicle introduction definitely will create environmental impact, traffic congestion, traffic congestion, accidents. Industries need to reduce these negative issues and to enhance aspects such as excitement, delight, and comfort (Figure 4.3).

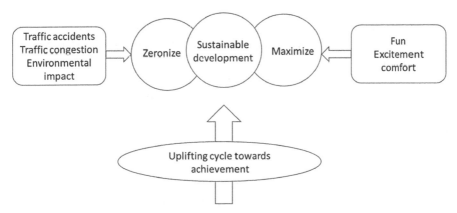

FIGURE 4.3   Zeronize and maximize (vision and philosophy) [1].

Toyota scientists have established technology revolution that brings about improved physical properties of recycled EPDM by selectively breaking the cross-links of three-dimensional EPDM network. Three factors are of more concerns of automotive industries-carbon dioxide emission, disposal of automobiles and air pollution from exhaust emissions. With introduction of Government, regulations for sustainability automotive industries are aiming for major changes in vehicle design, from Combustible Engine type vehicle to Electric Vehicle. By 2035, almost 90% of vehicle will be of electric type across the globe. To comply with future sustainability tire industries Research and Technology team goal is to develop new tire through design, manufacturing practice and introduction of sustainable material. Many new concepts are already under development like self-inflating tires, airless tires or Tweel tires, self-healing tires, 3D printed tires, tall, and narrow tires, triple tube tires, BOH_3 concept tire (to produce electricity while driving that would fill the battery) [12–16]. E-mobility trend in automotive sector means higher weight of vehicle contributed by battery. Higher weight of vehicles creates higher torque so electric engines need more efficient tire with higher wear resistance. For saving 1% fuel from fully loaded truck, only 2% to 4% rolling resistance reduction of the tire is sufficient, whereas passenger car tire requires 5–7% rolling resistance reduction to achieve same level of fuel economy [17]. The tire industry is raw material intensive which means that raw materials account for 65% of the production cost. Main raw materials for the tire are rubber (natural and synthetic), carbon black, silica, petroleum-based rubber process oil, nylon, polyester, steel cord, bead wire, and other rubber chemicals like resins, zinc oxide (ZnO), stearic acid, sulfur, etc. Raw materials used in the tire may be classified into two categories-from non-petroleum origin like NR, silica, steel cord, bead wire. Another raw material class is from petroleum origin like synthetic rubber, carbon black, rubber process oil, synthetic fibers like nylon, polyester. Table 4.1 shows the proportion of all raw materials in terms of their value and weight [18]. Most of the tire raw materials are petroleum-based. Hence, for tire companies are always getting influenced by raw material price because of the dependency of fluctuations of international prices of petroleum products.

Global tire chemical suppliers and mainly synthetic rubber manufacturers doing continuous research with non-fossil-based fuel materials to bring down the petroleum-based raw material usage by 30%. Synthetic Rubber manufacturers are trying to produce latest generation functionalized solution styrene-butadiene rubber (S-SBR) to facilitate silica mixing which will reduce carbon black usage. Industries are taking a new pathway, i.e., Styrene Circular Solution. This is nothing but recycling of polystyrene (PS) back into new styrene monomer for future usage [19].

TABLE 4.1 Proportion of Raw Materials in Terms of Value and Weight

Raw Materials	By Value (%)	By Weight (%)
Polymer	52	49
Reinforced filler	23	10
Petroleum-based tire cord	8	24
Rubber chemicals	15	12
Others	2	5
Total	**100**	**100**

4.2 POLYMER

Tire industry is raw material intensive and which accounts for 60% of the production cost. Most of the raw materials used in tire are petroleum-based except NR, bead wire, steel cord and few chemicals like stearic acid hence as we have spoken earlier that tire companies will be affected by the fluctuation of petroleum price. NR obtained from *Hevea brasiliensis* is the mostly used elastomer in rubber industries. Tire industries alone consume almost 75% of NR produced across the globe (Figure 4.4).

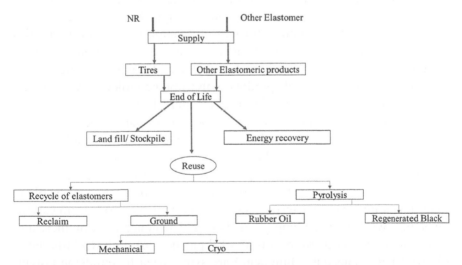

FIGURE 4.4 Flowchart showing the life cycle of elastomeric products.

To combat with the demand and price of NR, similar materials has been developed, e.g., isoprene or synthetic polyisoprene. They are well introduced and being used by all rubber industries. Because of the recent volatility in petroleum price, synthetic polyisoprene is also facing challenges to meet cost compatibility of industries. Scientists have found out other natural sources of NR, e.g., Guayule (*Parthenium argentatum*), which mostly grow in Mexico and Chihuahuan deserts of southern Texas [20], Dandelion (*Taraxacum kok-saghyz*) native to Kazakhstan of former the Soviet Union. Russian Dandelion is discovered by USSR during 1932 as a part of strategic programme of harvesting domestic source of NR [21]. It uses resources that resemble the one from a frequently used *Hevea brasiliensis* rubber tree. Different Research organizations across the Europe and the US are working to enhance the yield of Dandelion rubber approximately 2000 kg per hector [21]. The above research came into picture with the hope to create an alternative to Asia's rubber monopoly. There are other advantages for these two alternatives of NR, beside the monopoly issue. *Hevea brasiliensis* is vulnerable to several pests and its growth depends on tropical climatic conditions. Besides this, guayule and dandelion do not cause allergic reaction which is very common for *Hevea* rubber. The Neiker-Technalia center has been commissioned in Spain to research genotyping of the Guayule (a type of shrub) and Russian dandelion (a perennial plant) [22]. The former is cultivated in Mediterranean areas while the latter thrives in northern and eastern European countries. Guayule rubber is linear in structure and its crystallization rate is faster than NR, hence gum guayule rubber exhibits improved failure properties [23]. Resin content is less than 5% hence oxidative degradation properties are better than NR [24]. Apart from Hevea, Guayule, and Dandelion, other plants are also known which produce polyisoprene-Goldenrod (*Solidago altissima*), Jelutong (*Dyeraostulata*) [25] (Figure 4.5).

Sarkar et al. [25] in his review have mentioned many types of terpene based sustainable elastomers which are under development and few are already commercialized for industry uses. Elementary chemical structure of most of the terpenes are of isoprene type (2-methyl-1,4 butadiene, C5, "isoprene rule"). β-myrcene (7-methyl-3-methyl-1,6-octadiene) is one of the important monomers of the terpene family. It is available in oils from various plants like wild thyme, ylang-ylang bay, and lemongrass. Commercially β-myrcene is produced by pyrolysis of β-pinene [26]. Poly (β-farnesene), poly (ocimene), and poly (limonene) are synthesized by emulsion polymerization. Recent synthesis report of poly (styrene-co-myrcene) has found that its chemical structure is similar to SBR [27]. β-myrcene structure is similar to 1,3-polybutadiene, so this monoterpene is used for many chemical synthesis [25]. β-myrcene is usually found in liquid stage, easy for storage.

Although liquid handling in industry is a challenge but it makes the process greener and truly sustainable solution.

FIGURE 4.5 Various rubber plants: (a) Hevea, (b) Russian dandelion, and (c) Guayule

Source: Reproduced from Rubber Asia.

Newmark and Majumdar [28] have done the synthesis of poly (β-farnesene). Synthesis route that has been followed is anionic polymerization technique. The product's main constituents are of 1,4-cis microstructure. Sumitomo Rubber has introduced liquid Farnesene rubber very recently. This Farnesene rubber can result in better ice-grip properties for its new stud-less winter tire [29]. This proprietary liquid Farnesene rubber is developed by Kuraray Co. Ltd, Tokyo in collaboration with Amyris Inc.

Abdelrahman et al. [30] have reported catalytic ring-opening dehydration of tetrahydrofuran to produce 1,3-butadiene. Azeem et al. [31] have synthesized a styrene monomer from *Penicillium expansum* which is cultivated on forest waste biomass in the presence of yeast extract broth. Styrene biosynthesis from sugar over *E. coli* strains has been invented by Mckenna and Nielson [32].

Amyris and other companies have aimed for large scale production of terpenes via fermentation of plant sugars and cellulosic wastes [33–35]. Amyris has modified the common yeast *Saccharomyces cerevisial* to efficiently produce trans-β-farnesene [36]. Global energies in partnership with Synthos are developing a fermentation process to produce butadiene using modified *Escherichia coli* [37]. Cattle hair keratin is developed from tanning industry wastes and hydrolysate of bird feathers from poultry wastes [38]. Barone et al. [39–41] have explained that bird feather keratin has a 40% hydrophilic part and 60% hydrophobic part that enable it to react easily with hydrophobic polymer and hence improving its strength parameter. Prazepiorkowska [42] in his review have mentioned that elastic polymeric films containing keratin can be used as motor car parts. Keratin incorporation in NR and nitrile rubber improves thermal aging resistance of the matrix. Keratin contains sulfide bonds, which during aging are converted into groups with anti-oxidant character.

The BF Goodrich Co. has manufactured passenger car tires and the United States Rubber Co. Ltd has produced heavy-duty rayon cord truck tires from Dandelion rubber [43]. Road tests reveal that the tread wear, of dandelion, made tires are 94% as good as that of guayule rubber. Building tack of dandelion is better than guayule and similar to that of NR. Soybean oil used in Goodyear Tire can increase tread life by 10%. In 2012, Bridgestone have announced a tire made out of soybean oil [44]. Goodyear and DuPont Industrial Biosciences are working together to develop bio isoprene. Pirelli and Versalis have initiated joint research project for the use of guayule based NR for tire production [45]. In 2012, Dutch Tire Company Apollo Vredestein have produced prototype of tires using NR made from Guayule and Russian Dandelion. In 2015, Bridgestone Japan produced tires using Guayule rubber. Michelin has announced that by 2048, all of its tires will be manufactured using 80% sustainable materials and 100% of all tires will be recycled. Currently, they are using 28% sustainable materials like NR, sunflower oil, limonene, etc., and 2% recycled materials like crumb rubber [36]. The Bio Butterfly program is launched in 2012 with Axens and IFP energies Novelties, to produce synthetic elastomers from biomass like wood, straw, beat. Michelin and Axens have announced that first industrial scale prototype plant will start by end of this year 2019 for manufacturing butadiene from bioethanol [46]. Falken's clever new Enasave 100 tire has been made using natural alternatives replacing synthetic rubber, and conventional carbon black with biomass-derived carbon black. Instead of using fossil resources for rubber antioxidants, the tire is made by a unique process. They have used plant oil instead of mineral oil with NR and silica. Obviously, these tires have eco-friendly benefits. Enasave 100 tires consist of better wear resistance, better fuel efficiency, and wet grip performance. Michelin are aiming for replacing oil-based elastomers by wood chips from wood industry waste. They are hoping to have first wooden tire sometime in future in 2020 [47]. Bridgestone have displayed concept tire of 100% sustainable material at the 2012 Paris Motor show and commitment is that by 2050, 100% sustainable material usage. Bridgestone is hoping for, by the year 2020 commercial sales of sustainable tire will take place [48].

Epoxidized natural rubber (ENR) is capable of substituting the expensive solution polymerized styrene butadiene (S-SBR) rubber which is widely used for high performance car tires, including winter tires. Truck bus radial (TBR) tire compound formulations are mostly based on NR along with carbon black as reinforcement filler to cater mechanical and wear properties. Replacement of carbon black from TBR formulation is a challenge since silica incorporation as reinforcing filler in NR is difficult. So, to minimize the usage of carbon black ENR along with NR is being practiced which facilitate silica incorporation in TBR tire. Recently Apollo Tires and Rubber Research Institute of India (RRI) partnership has patented special grade of ENR.

Solution styrene butadiene (S-SBR) is one of the key elastomers used in tire apart from emulsion styrene butadiene rubber (E-SBR) and polybutadiene rubber. S-SBR plays an important role for low rolling resistance in combination of silica filler. Precise molecular design for controlled microstructure and chain end structure with special functional groups are pursued [49]. Such functional components are amino, carboxyl, and epoxy groups (Table 4.2).

TABLE 4.2 Possible Molecular Design for Controlled Microstructure and Chain End Structure [49]

	M_W	ML_{1+4} 100°C	Resilience
	35.7	56	57
	34.8	51	64
	37.7	63	62
	29.5	53	65

Materials development trends are moving in two distinct directions—Collaborative research with polymer producers on tailored synthetic elastomers for niche applications and development of sustainable alternatives to traditional petrochemical-based materials. Researcher has developed S-SBR with functional groups at one chain end, both chain end and also in backbone chain. Different degree and type of functionalization provide better silica-rubber interaction for improved snow traction, wet traction, and rolling resistance properties of ultra-high-performance passenger radial tires. Tire pressure is one of the important criteria to be maintained by the user to get the maximum benefit of fuel economy. The tire component responsible for retaining tire pressure is called inner liner. Butyl and halo butyl rubber are targeted for substitution by fossil-free material. Dynamically vulcanized rubber/resin thermoplastic elastomers alloy (DVA: dynamically vulcanized alloy) is therefore evolved. DVA is a sea-island structure, with a resin as sea phase and crosslinked rubber as island phase [49]. The resin has some barrier properties. A DVA-based inner liner can support both high barrier performance and endurance. This information is unveiled at the 2009 Tokyo motor show as next generation eco-tire technology.

Continental launches bike tire made from sustainable dandelion rubber [50]. Goodyear's Fuel Max Technology features a proprietary fuel saving tread compound. Hankook Tire Corporation has introduced its future generation electric vehicle tire which features strengthened handling and noise reduction capabilities. It provides hyper low noise environment, comfortable riding, and driving performance similar to ultrahigh performance tire. Aramid hybrid reinforcement belt features heavier load carrying capacities. This tire also contains aqua pine compound, resins extracted from conifers and vegetable oil. The new compound exhibits better wet braking and driving stability. This new generation treads compound offers improved tread abrasion resistance to combat high power of the electric vehicle and initial acceleration with no loss [51].

4.3   RECYCLING MATERIALS

Uses of rubber products in day to day life are increasing with time. Disposal of used end life rubber products is of serious concern. Tire is one of the rubber products which is contributing major stockpile up. Approximately 73% of scrap tires are disposed of in landfill [52]. Goodyear Tire has achieved their goal of not sending any waste to landfill from any of their manufacturing facilities and in 2006; they achieved zero waste to landfill [51]. Scrap tire dumping to the land continues to pose serious environmental, health, aesthetic problems as they are bulky, non-biodegradable, good breeding area for mosquitoes and rodents, shelter for animals like snake, rats, and also high risk of fire. Moreover, many toxic chemicals and process oils leaching to the soil is detrimental for fertility of the land as well as pollution of underground water supplies. Incineration of tire as a supplement of coal, wood can again multiply the environmental hazards. Reuse of scrap tire is thus a burning issue on which scientists are working to find out solution to maintain sustainability by converting it into a secondary resource of raw material for industry use. Recycling is one of the solutions which are practiced more than two decades. That we are trying to discuss in this chapter.

The U.S. Tire Manufacturing Association (USTMA) released its "2017 U.S. Scrap Tire Management Summary" report in July 18, 2018. The report reveals that in 1991 over 1 billion scrap tires are in stockpiles, now it's 60 million. So, there is 94% decrease [53]. This improvement has come due to the enforcement of stringent regulations. According to report of the European Tire and Rubber Manufacturer's Association (ETRMA), the treatment route of end-life-tire has changed from land-filling to material and energy recovery regime. Distribution report of scrap tire disposition in the U.S. in 2017 is shown in Table 4.3 [54].

TABLE 4.3   Scrap Tire Disposition in the U.S. in 2017

Disposal Category	Percent (%)
Tire derived fuel	43
Ground tire rubber	25
Land disposed	16
Civil engineering application	8
Exported	3
Electric furnace fuel	1
Reclaim project	1
Others	3

Some of the ways of using end-life-tire include; use in asphaltic concrete, for paving streets and highways, sport surfacing, production of steam through incineration, as fuel for cement kiln. In the tire industry, it is used after converting into recycled material like rubber crumb of finer mesh, chemically treated, or de-vulcanized rubber as secondary raw material. As of 1990, established tire and rubber industries are using only around 2% recycled material. However, in the past decades the tire recycling industry

has a tremendous growth. Coupled with legal restrictions and environmental pressure present consumptions of recycled materials is nearly
10%. Recycling means conversion of used rubber into a re-processable
and reusable form by the currently used physical and chemical process
to save resource, energy, and to control the spread of hazardous/non-
biodegradable materials [1]. The converted new product is treated as a
secondary source of raw material for use in tire and other rubber industry.
Reversing the vulcanization process is sometimes compared with reversing
breadcrumb into fresh dough or like un-baking a cake and then reusing
the eggs [1, 55]. In chemical term, de-vulcanization means reversing the
rubber from its thermoset state to elastic state by cleaving sulfur bonds
in the molecular structure [55]. The recovery of useful, uncontaminated
materials from tire is very complex process. Scrap vulcanized rubber
utilization is first thought in 1858 when Hiram Hall have developed heater
pan process for reclaiming NR vulcanizates [52]. The major component of
steel and fabric free scrap reclaim rubber is-rubber (50%), carbon black
(27.5%), oil (17.5%), and ash (5–10%) [56]. The heating value of scrap
tire is approximately 35×10^6 Joule/Kg [52]. Chemical analysis of whole
scrap tire reveals carbon – 83%, hydrogen – 7%, oxygen – 2.5%, sulfur –
1.2%, nitrogen – 0.3% [56]. In 1990 Watson have used a novel mixer to
explain that in recycling process C-S bonds can rupture rather than C-C
bonds, since C-S bond strength is less than C-C bond [57]. A high strength
mixer (HSM) is presented in 2009 as an efficient reclaiming machine [57].
In this process mechanical strain and chemical reaction occurs to rupture
cross-linked bond and polymer backbone remain unaffected. ASTM 184A
defines devulcanization as a "combination of depolymerization, oxidation,
and increased plasticity. In other words, devulcanization is the process of
cleaving of monosulfide, disulfide, and polysulfide crosslinks of rubber. In
comparison to this process, reclaiming is different to from devulcanization
since rupture of the C-C bonds also happen in this process.

The major form of recycled rubber is still ground rubber. This is produced
either by ambient grinding, wet grinding, and cryogenic grinding. The
product outcome of this process is called rubber powder or crumb rubber
which may be of different types of coarse or finer grades depending on mesh
size used for making crumb. Out of the above three techniques cryogenic
grinding provides better quality tire crumb rubber and it is more economical
and safer than ambient grinding [58] (Table 4.4).

Recently Michelin have acquired Lehigh, a specialized in high
technology rubber powders which derived from recycled tire. Michelin

proposes to develop partnership and identify new ways to recycle tires or new outlets for recycled tires [48]. Recycling of rubber can be followed by different techniques like mechanical, chemical, thermo-mechanical, cryogenic, microwave, and ultrasonic process. The ultrasonic reclaiming process has been explored in 1973 by immersing rubber articles in a liquid followed by treatment with ultrasonic radiation in the range of 20 KHz26/79. It is reported that by this process C-S and S-S bonds cleavage occurs but the C-C bonds remain intact. The properties of this recycled rubber are found to be similar those of original vulcanizates. Aoudia et al. [60] have reported microwave devulcanization of waste tire; the same has been used for making epoxy composites of improved mechanical properties.

TABLE 4.4   Pyrolysis Products of Used Tires [59]

Primary Products	Wt (%)	Content	Products Derived from Primary Products
Gas	10–30	Hydrogen, CO_2, CO, methane, ethane, propane, butane, sulfur, and other hydrocarbons	
Oil	38–55		Carbon Black
Char	33–38		Activated Carbon

Chemical process for reclaiming is more effective and faster devulcanization is selective for crosslinking only. Mercaptans, disulfide, phenol, amines, and metal chlorides are basic chemicals used for this method. Studies on Mechanochemical recycling processes of NR based scrap rubber indicate that diallyl disulfide (DADS) played an important role in the process [61, 62]. Diphenyl disulfide is found to be an effective reclaiming agent for NR based scrap rubber but it has bad odor [63–65]. Kojima et al. [66] have used supercritical carbon dioxide and diphenyl disulfide for chemical recycling of NR based scrap rubber. TMTD has been introduced as an alternative reclaiming agent of DADS by De et al. [67–69]. Reclaim prepared by using TMTD as reclaiming agent provides better tensile properties when mixed with SBR compounds. Verbrugger et al. [70] have used various types of amines as reclaiming agent for EPDM based scrape rubber compound and explained that there is no difference reactivity in different types of amines. Yehia et al. [71] studied the reclamation reaction by phenyl hydrazine system in brabender mixer along with ferrous chloride, ferric chloride and zinc chloride. The use of alkaline metals to break the crosslinks in rubber was patented in 1997 [72]. In this process, devulcanization has been done in solvent like, toluene, benzene, naphtha in the presence of sodium.

Surface modification method for making ground rubber has been done without breaking the bonds in the vulcanized material. Abrasion resistance and tear strength properties are influenced by surface modification. The most common method for chemical activation is treatment of powder by cross-linkable polymer-like latex, low molecular weight polymer and a curing system. The additives are latex, epoxidized oligomers, acryl amide. Grafting of ground tire rubber with maleic anhydride, methyl methacrylate, ethylene vinyl acetate, acrylamide has been reported [73–75]. Extensive work has been done on anaerobic desulfurization by the microorganisms *Pyrococcus furiosus*. This bacterium converts sulfur to hydrogen sulfide [76]. Another bacterium *Sulfolobus acidocaldarious* can split C-S bond. It is oxidizing sulfur but the process can be stopped at intermediate stages [77, 78]. Pyrolysis is another route for conversion of scrap tire into useful products like TDF and making regenerated carbon black [79, 80]. Pyrolysis is usually followed through different techniques like in inert atmosphere, oxidative pyrolysis, steam atmosphere, molten salt technique. Manuel and Dierkes[81] review on pyrolysis development in Europe and Japan. In general pyrolysis products are 5–20% light hydrocarbon gas, 20–50% hydrocarbon oil and aromatic liquids and 30–50% carbon char. Glenn and Ward studied the use of ground scrap rubber to prevent weed growth particularly on soybeanfields [82]. When crumb powder is used as mulch soybean plants are reduced in size thus minimizing wind damage, no adverse effect on seed quality. Mandal et al. [83] in his research has worked on regenerated CB from tire scrap, explained that regenerated CB has higher surface area (due to increased surface roughness) than virgin black which resulted better aging properties than the virgin black.

Bandyopadhyay et al. [84] have studied tire crumb use in NR based tire tread cap compound. He observed marginal deterioration in tensile strength, fatigue to failure and abrasion properties. 100 mesh crumbs have shown better properties retention compared to 40 and 80 mesh crumbs. In another study Bandyopadhyay et al. have explained that incorporation of superfine reclaims in the NR-based tire tread compound retained the properties to a greater extent compared to other recycled product like devulcanized rubber and other coarser reclaim. Regenerated carbon black from scrap tire compound is studied in NR based Lug tire tread compound and compared with virgin N330 black. Accelerated aging properties of compounds containing regenerated black have been found to be marginally better than that of compounds containing N330 black [85].

4.4 SUSTAINABLE RUBBER ADDITIVES

Elastomers are very often used as its normal form. For the manufacturing of finished product, a series of compounding ingredients are getting along with the base elastomeric material. For the ease of understanding, a typical tire tread formulation with all the possible rubber additives has been given in Table 4.5. Each and every compounding ingredient has their specific roles and responsibilities. That's make them inescapable from the product formulation and compounding. Most of these rubber chemicals are from inorganic sources or from petrochemical sources. Hence, mostly all are associated with health hazard and have a huge impact on environment. One solution to get rid of such problems can be the full substitution of these chemicals without disturbing the properties that those chemicals indulge into the finished product. But after an extensive search, a complete substitution is not at all possible. So, on a second thought, one can minimize the risk associated with those chemicals. And that can be achieved by following a sustainable pathway. Thinking about the market, scientific communities have put a very careful and slow-moving way forward towards the plausible alternatives of these rubber chemicals (Figure 4.6).

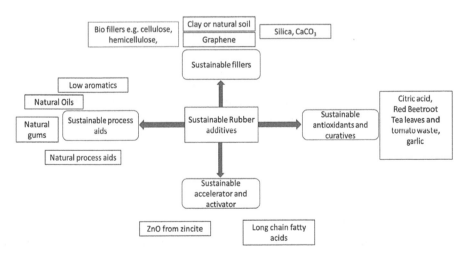

FIGURE 4.6   Schematic representation of various sustainable rubber additives.

In this section, we are trying to have a short discussion on these additives from a sustainable viewpoint. It includes sustainable rubber additives and curatives, various process aids such as vegetable oil and natural gums, sustainable fillers such as clay, calcium carbonate, silica, graphene, bio fillers from plant wastes, sustainable steel cord, etc. That also includes some possible sustainable alternatives.

TABLE 4.5 Typical Tire Tread Formulation (phr: Parts Per Hundred Parts of Rubber)

Ingredients	Amount (phr)
Natural rubber	100
Carbon black	45
Low PCA Oil	10
Stearic acid	2
ZnO	4
Antioxidant 6PPD	2
Antioxidant TMQ	1.5
Accelerator	1.5
Prevulcanization inhibitor (PVI)	0.2
Soluble sulfur	2.0
NOBS	0.5

4.4.1 ECO-FRIENDLY PROCESS OILS AND VEGETABLE OILS

Process oil is an important tool, which is being added during the processing step for the incorporation of filler materials within the rubber matrix. It affects the properties of the rubber vulcanizates from hardness and modulus point of view. With a growing concern for environmental safety, there is a global demand for adopting a green concept or a sustainable pathway. Depending on the source of the material and their chemical composition, "oil" can be classified. For composite application, use of oil is mostly as a process aid which is nothing but an extender. In case of automotive application such as for tires, high aromatic oils are most commonly used. High aromatic oil is well compatible with a wide range of synthetic and NR. But its source lies on the petroleum crude only. These oils are having a high content of polycyclic aromatics (PCA) which are potentially carcinogenic. From the wear of a tire tread PCA is getting leached into the environment since they are not chemically attached to base rubber material. This environmental effect leads to a global need to reduce the dependence of petroleum-based oils. So, there is a need to aware the user to promote the use of low PCA and low polyaromatic hydrocarbons (PAH) containing process oils in place of using old conventional aromatic and naphthenic process oils. PCA levels upper than 3% are alarming for the environment, according to REACH regulations (Figures 4.7 and 4.8).

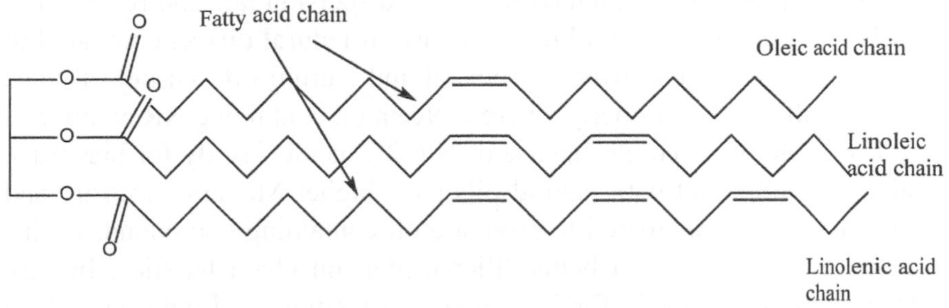

FIGURE 4.7   Chemical structure of vegetable oil (in general).

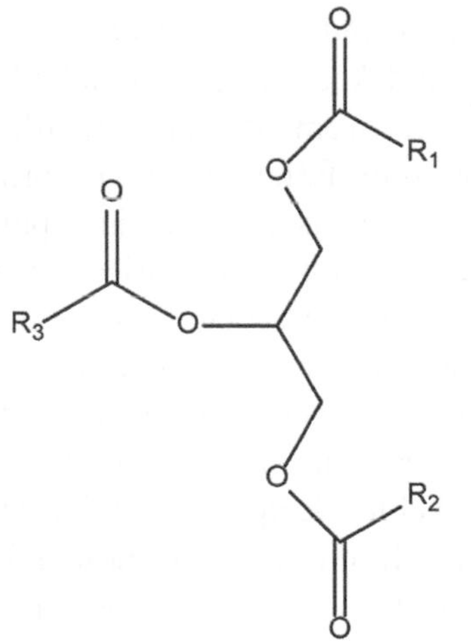

FIGURE 4.8   Chemical structure of triglycerides where R1, R2, and R3 are the fatty acid units.

The best alternative for petroleum-based resin can be vegetable oil. Vegetable oil belongs to such a platform where it is inherently biodegradable, extremely low cost and having excellent credentials towards environment. Figures 4.7 and 4.8 shows the general structure of triglycerides which is nothing but the general chemical formula for vegetable oil. Their tautomeric structure possesses unsaturation that can polymerize in the presence of atmospheric oxygen.

Dasgupta et al. [86–89] have studied the effect of natural source oil as an alternative to low PCA oils for rubber industries. They have compared the effect of Neem oil and Kurunj oil with aromatic oil on 100%NR, NR/BR blend, and SSBR/NR/BR ternary blend. And they have observed some interesting facts. For 100% NR compound, aromatic oil, Neem oil, and Kurunj oil shows similar flow characteristics. Filler dispersion rate and retention of modulus after aging is much higher in case of natural oils as compared to aromatic oil. Again in between Neem oil and Kurunj oil, Kurunj oil have better filler dispersion property whereas Neem oil containing NR compound have lower Mooney scorch. In case of NR/BR blend (mostly for bias truck tread cap), compound with natural oils have higher Mooney viscosity and higher abrasion as compared to aromatic oil containing compound. In this case, neem oil has shown better filler dispersion characteristics. In case of ternary system (mostly for PCR tread cap), compound with natural oil shows higher polymer to filler interaction. They also show higher modulus but tensile strength retention is pretty poor as compared to aromatic oil. Dasgupta's group have also studied the effect of aneco-processing aid and zinc soap-based process aid known as Zincolet PN60 with SBR/NR/BR ternary system. It shows improvement in polymer filler interaction that leads to better mixing, heat buildup, fatigue to failure properties. Vegetable oils such as linseed oil are long time used for coating application mostly as paint and varnishes. Vegetable oil can be easily dried. Castor oil and soybean oil are another typical examples of vegetable oil which has very good color retention property. Vegetable oil can be used as a resinous plasticizer as well as heat stabilizer. Typical resinous plasticizers are cottonseed oil or coconut oil. Common example of heat stabilizer oil is rubber seed oil which imposes heat stability on polyvinyl chloride application. Heat treatment of fatty acids using sulfur and accelerator lead to vulcanization of the fatty acids. Botros et al. have studied the effect of vulcanized oils on rubber compound and how it affects flow characteristics and the aging properties [90–93]. Ismail et al. [94] have used palm oil in silica filled NR compounds and studied their properties. It appears to have better vulcanizate properties. In case EPDM rubber, Wang et al. [95] have studied the effect of palm oil in comparison to paraffinic oil. The effect of both the oils is comparable if we consider processing and mechanical properties. But due to low cost and natural source, palm oil can be a useful alternative of paraffinic oil in case of EPDM rubber. In another study epoxidized, palm oil comes in direct comparison to distillate aromatic extract oil. It has been observed that both the vulcanizate possess similar properties such as hardness, modulus, tensile strength, elongation at break, etc. These types of findings have placed epoxidized palm oil as a plausible alternative for distillate aromatic extract oil [96–98]. Even for the carbon black containing butyl rubber compound, the use of norbornylized soybean oil has been reported [99]. Linseed oil have found its use as plasticizer in rubber/EG matrix. It appears to have higher thermal stability, resistance to crack formation as compared to the naphthenic oil [100].

4.4.2 SUSTAINABLE ACCELERATOR AND ACTIVATORS

ZnO is an extensively used material in rubber industries because of its role it plays in rubber vulcanization process. It acts as an accelerator activator in combination with stearic acid. ZnO is a multifunctional material based on its unique physical and chemical properties. The preparation of ZnO involves metallurgical or chemical method. Chemical method of preparing ZnO is nothing but the emulsion and microemulsion technique. And metallurgical process includes mechanochemical, sol-gel, hydrothermal, and solvothermal process directly from Zincite [101]. Because of its extensive use, Tire industries are the biggest consumer of ZnO. They consumed almost a half of the total worldwide production of ZnO [102]. Commonly ZnO is used in tire formulations in 3–8 phr (parts per hundred parts of rubber) (Table 4.5). The growing concern for ZnO is its environmental impact. ZnO always comes with a small fraction of cadmium oxide (CdO) and lead oxide (PbO) and this leads to a proper REACH regulation of controlling the amount of this trace element (heavy metals). On the other hand, stearic acid is a long chain fatty acid with a C18 chain. Since both ZnO and stearic acid originates from natural source, so from sustainability point of view one can consider both these materials as green material.

A continuous effort is ongoing to reduce the amount of ZnO to get rid of the environmental issues and also to lower the production cost of tire. Out of different approaches, one solution can be the use of nano ZnO in place of normal ZnO. Nano ZnO possesses higher surface area and that significantly affect the mixing characteristics. The dispersion of high surface area nano ZnO is better than the normal rubber grade ZnO and that allows using ZnO at a lower level with much more confidence [103]. Sahoo and Bhowmick [104] have studied the role of nano ZnO in XNBR compound and reported that ZnO nanoparticles induces higher state of curing and higher maximum torque as compared to rubber grade ZnO. In another part of the research, people have worked to find out better alternative of ZnO. Magnesium oxide (MgO) is found to be the most promising metal oxide out of the block. Being a non-heavy metal oxide, it insists the breakdown of the accelerator during the sulfur vulcanization process. And the rate appears to be faster than ZnO. The main drawback of using MgO is the lower percentage of crosslinking that actually hinders industrial application.

To overcome such issues some new technique has been introduced. Przybyszewaska et al. [105] have studied that in case of NBR, zinc chelates can be used in the vulcanization process. Guzman et al. [106] have reported the development of a new activator which consist a mixture of both ZnO and MgO. In this accelerator, Mg is incorporated inside the ZnO lattice with a very precise stoichiometry to form mixed metal oxide $Zn_{1-x}Mg_xO$ in nano scale. In the sulfur vulcanization step, it has been observed that this mixed metal oxide takes advantages from both MgO and ZnO. The reactions that generally happen during the scorch time, the chemical decay of the accelerator and the formation of the MBT happens much faster than it was expected earlier. And that happens due to the presence of Mg inside the ZnO lattice structure. We have already discussed that the low crosslink density is one of the major drawbacks of using MgO alone. That problem can be resolved by using this mixed nano oxides.

4.4.3   NATURAL GUM AND OTHER MULTIFUNCTIONAL PROCESS AIDS

The best source of natural gum is seed coatings or woody parts of plants. It is nothing but utilizing waste materials. So, it has gained potential interest from environmental aspect. A lot of advancement has been done on this field. Earlier very, less attention has been paid for utilizing such waste in rubber formulation. One of the eco-friendly approaches is by utilizing waste natural gum in virgin rubber.

In recent times Bahera gum have received a lot of attention from the rubber fraternity. It comes out from the bark of a medicinal plant, called *Terminalia bellerica*. This tree can be easily found in south East Asia, mostly in India. The extract of this tree is rich in polyphenolic compounds, which makes it useful as a chemo preventive agent. Except that Bahera gum is full of different tannins and calcium oxalate [107–109]. The gum is basically a multifunctional additive. The fatty acid/esters present in the gum can serve as accelerator activator whereas the phenolic part acts as antioxidant. Guhathakurta et al. [110] have used Bahera gum in NR and brominated iso-butylene-co-paramethyl styrene rubber. Addition of the gum in Rubber increases the tack strength. Activation energy has also been decreased with higher loading of gum. Saha et al. [111] have used Bahera gum in water-based NR formulation. Lap-shear strength and peel strength are found to be increased by a huge margin. Shanmugharaj et al. [112] have studied the influence of Asan gum in water-based NR latex system. At an optimum 5 phr of gum loading, they have observed around 400% increases in lap tear strength for NR latex system. The same system has been used for poly (vinyl acetate) and poly (vinyl alcohol) mixture. At an optimal 5 phr loading there is around 15% increase in lap tear strength for the PVA system.

Another vulcanized animal/vegetable oil (can also be called natural gum) which is never been in limelight is Factice. It can be used as a polishing ingredient for finished rubber product. Factice possess all the minimum characteristics it can have to act a processing aid or a property enhancer for rubber compounds. Two main classes of factice are heat-cured brown factice and cold cured white factice. They are introduced into the rubber field almost at the same time [113].

Rubber seed oil is another kind of natural gum that is extensively getting used as a multipurpose reagent in NR and SBR compounds. Nandanan et al. [114] have studied that. From one angle rubber seed oil can enhance the mechanical properties and from the other side it can enhance abrasion, flex resistance and aging resistance. Rubber seed oil is also helpful for decreasing cure time and blooming on the vulcanizate surface. There are other different resin, promoters, and some supporting literature also exist [115, 116].

4.4.4 ANTIOXIDANTS AND CURATIVES

Antioxidant and curatives are added into the rubber formulation in a lesser amount as compared to other ingredients. But from sustainability point of view, their replacement can certainly bring something positive for the environment. Here is this study we are trying to present some of the benign source for these antioxidants and curatives that are useful for rubber compounding formulation.

Tomato fruit is an important source of antioxidants. Pinela et al. [117] have studied that. It requires some extraction process that can generate those antioxidants from tomato. It is proven that a lot of phytochemicals and phenolics such as catechin, quercetin, cyanidin, coumarins, etc. have a huge application potential as a benign antioxidant. Shahidi et al. [118] have reported the ease availability of antioxidants from different nuts. Wu et al. [119] have done something interesting with citric acid. From citric acid, through a microwave pyrolysis technique, carbon nanodots are generated. These carbon nanodots are none other than antioxidants which are not radical friendly. Wu and group have used these nanodots while mixing with SBR. Antioxidants can easily be termed as stabilizers for several polymer systems [120].

Tran et al. [121] have worked with red beetroot which contain a specific antioxidant called betanin. They have incorporated this beetroot powder into a starch-based elastomer to obtain flexible bio-composites with tunable antioxidant properties. This smart bio elastomer composite can find its application in the field of pharmaceutical and food packaging. Most of the thiuram, those are used as accelerators for rubber industries are not environment friendly. They release carcinogenic nitrosamine during the heating. TBzTD (tetrabezylthiuram disulfide) has come to the market with the prospect of minimizing risk of using conventional thiurams. Metal salt of sorbic acid and ferulic acid has been introduced by Lin and his group [122]. Zinc salt of these acids is found to be a suitable crosslinking agent for ENR/silica composites. There has been noticeable increase of abrasion resistance in case of zinc salt of ferulic acid. Garlic and related genus allium plants are a good source of organic sulfur. This diallyl disulfide can be strong contender for replacing common sulfur in rubber compounding [123, 124].

4.4.5 SUSTAINABLE FILLERS

Virgin elastomer does not have load-bearing capability. It does not have that sufficient strength. So, it is necessary to add fillers into it. It can be carbonaceous (carbon black, CNT, and graphene) or silica, clay, or other renewable bio fibers such as cellulose, hemicellulose, or lignin. In this part of the study, we are trying to present some of the renewable fillers from sustainability point of view (Figure 4.9).

Introducing nanofillers into the tire industries is also something related to sustainability. By introducing 2–4 phr of nanofiller, someone can achieve same level of properties as it is for the conventional filler loading. That can reduce the weight of the final product, i.e., tire for automobile industries. So, it has a direct/indirect impact on fuel consumption.

4.4.5.1 SILICA

Silica can be of two types; one is fumed silica and another is precipitated silica. Precipitated silica is most common for rubber and tire industries.

Silicon dioxide, also known as silica (from the Latin silex), is a chemical compound that is a dioxide of silicon with the chemical formula SiO_2. It has been known since ancient times. Silica is most commonly found in nature as quartz.

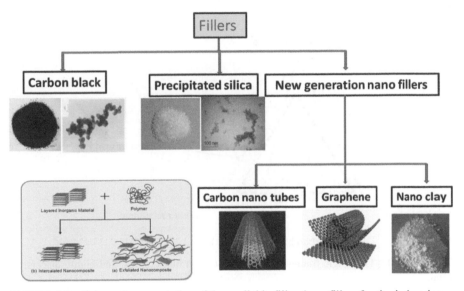

FIGURE 4.9 Schematic presentation of the available fillers/nanofillers for tire industries.

Addition of silica to a rubber compound offers a number of advantages such as improve tear strength, less heat buildup and increase in compound adhesion in multi-component products such as tires. Two fundamental properties of silica influence their use in rubber compounds: ultimate particle size and the extent of hydration. Properties such as pH, chemical composition, and oil absorption are having pretty less importance. Silica, an amorphous material consists of silicon and oxygen arranged in a tetrahedral structure of a three-dimensional lattice. Particle size ranges from 1 to 150 nm and surface area from 20 to 300 m^2/g. There is no long-range crystal order, only short-range ordered domains in a random arrangement with neighboring domains. Surface silanol concentration (silanol groups-Si-O-H) influences the degree of surface hydration. Silanol types fall into three categories; namely isolated, geminal (two -OH hydroxyl groups on the same silicon atom), and vicinal (on adjacent silicon atoms). Using new generation silane coupling agent, silica rubber interaction can be improved drastically. The formation of filler-rubber linkages via organosilanes has a major influence on the properties of rubber compounds. By lowering the specific component of surface energy of silica, the level of adsorption of hydrocarbon rubber on the filler surface can be enhanced and subsequently filler-filler interactions are reduced. Coupling agents are used as a surface modifier, either applied on the silica particle itself before the mixing or during mixing to the rubber compound [125–130] (Figure 4.10).

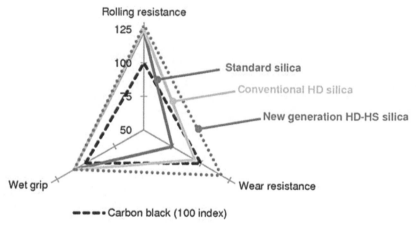

FIGURE 4.10   The magic triangle of tire with carbon black, silica, and highly dispersive silica.

Source: Reproduced from Ref. [131].

Michelin are the first company to introduce silica in a tire formulation to have their first-ever fuel-efficient green tire. Using silica as reinforcing filler, they are able to reduce tire rolling resistance by 20% and hence making the tire 3–4% more fuel economic. Also, silica tires are suitable for winter applications because of its good flexibility at low temperature [132]. Silica manufacturing giants such as PPG, Solvay, and Evonik are working to produce highly dispersive silica (HDS). This HDS can be one of the solutions of "Magic Triangle," that all tire industries are aiming for (Figure 4.10). Some typical examples of HDS are ZEOSIL, ULTRASIL, etc., [133–135]. These HDSs are leading to better dispersion that can affect the tire properties from three different angles. Higher rubber filler interaction can lead to better wear resistance. Lower filler-filler interaction can lead to lower hysteresis at high temp, hence lowering the rolling resistance [136]. Nayek et al. and Bhowmick et al. have reported the wear behavior of NR/SBR/silica composites. The idea of introducing green tire by varying silane and silica variant [137, 138].

From the earlier reviews, it has been estimated that the amount of biomass from plant resources is more or less 10^{13} tons worldwide and only 3% of it is getting renewed per annum. To maintain overall oxygen balance, the theory of sustainability restricts the usage only at the renewed percentage.

4.4.5.2 CELLULOSE

Introduction of cellulose opens a new era of polymer composite. Cellulose is a kind of renewable natural polymer which is most plentiful in the world. Plants are the source of cellulose as each and every plant cell wall is consisting of cellulose. From wood, grass to even seed fibers, fruit leaves, fungi, cellulose can be extracted [139–141].

Cellulose is basically a polymer of semicrystalline polysaccharide with a repeating unit of D-anhydroglucose. Cellulose consists of crystalline, paracrystalline, and amorphous domains [142]. Chemical structure of cellulose is illustrated in Figure 4.11 which is basically a cellobiose unit where n is the degree of polymerization. Degree of polymerization and hence the molar mass depends on the source. In case of wood fiber DP is around 300 whereas in case of plant fiber or bacterial cellulose DP is much higher, around 10000. Crystalline domain is responsible for the mechanical behavior of the cellulose fiber. Sakurada et al. calculated Young's modulus of elementary cellulose fibrils of bleached ramie fiber to be 134 GPa. Ishikawa et al. published a similar article where they measured the elastic modulus of different polymorphs sourced from ramie fiber. Their morphology and aspect ratio can vary greatly depending upon the source and also the polymorphs, so as their mechanical properties. The secondary hydroxyl groups at C2 and C3 that can form intra and inter molecular hydrogen bonds, play an important role in crystalline packing and hence physical properties of cellulose [143].

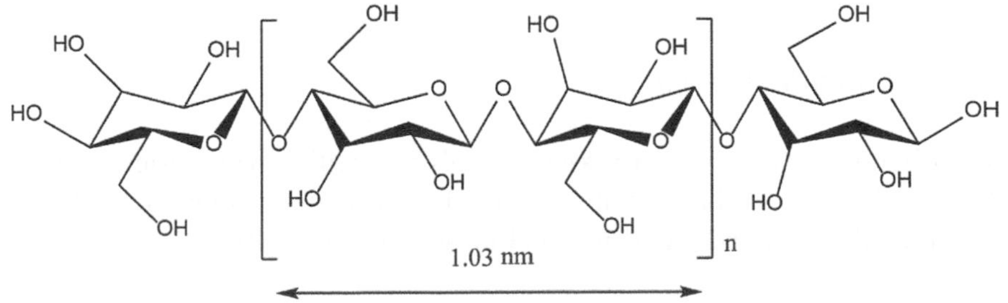

FIGURE 4.11 Chemical structure of cellulose fiber.

However, high-end application of cellulose is pretty limited due to lack of melting properties and its nature of being hygroscopic. The application of cellulose has been drastically increased only after the introduction of nanocellulose. Any cellulose microfibril, having a sideways measurable length below 50 nm can be considered as nanomaterials [144–146].

From 1970 onwards, the news of using natural fiber-reinforced composites start coming in Refs. [147, 148]. In the past, few years' nanotechnologies emerged as a unique technology with a very diverse role in the development of nanocomposites. It also reflects a constant growth in the market mostly for industrial applications [149]. Several groups are working on the development of nanocellulose fiber reinforced polymer composites. Compared to industrially made fiber and filler, nanocellulose fiber has gained much interest especially among automotive industry, in the sense of its availability and an improved CO_2 balance [150–157].

Agricultural biomass refers to the profusion of organic materials, so as to say the agricultural feed consisting of. It is the most abundant resource as well as it is renewable. Concerning it to be a fact, the society of fossil resources has been a precursor of the usage of agricultural biomass mostly in Southeast Asia and thus establishment of a sustainable production towards the usage of agricultural resources is pertinent [158]. Recent advances towards the biomass waste development and their conversion into products offer a huge opportunity for further research on improvised plant resources.

Cellulose fiber holds biorefinery functional properties towards agricultural waste and holds a good step towards the renewability of cellulose. With accordance to the assembled nanofibers are not more than a few micrometers in length. Because of the high gas barrier properties and their morphology, cellulose nanofiber stands out to be an optimistic material in various fields [158]. Some of the nanocellulose materials obtained from agricultural products or waste.

4.4.5.3 HEMICELLULOSE

Wood hemicellulose is also a kind of polymer with asymmetric polysaccharides. It has a similarity with the cellulose structure. Figure 4.12 shows one of the three characteristic hemicellulose microstructures.

Hemicellulose helps in improving some of the properties of cellulose used in paper industries. The rest of these polysaccharides remain attached with lignin and most of the processes it is getting burned with it. In the last few years new areas of using hemicellulose has been studied for several industrial sector in particular for food additive sectors.

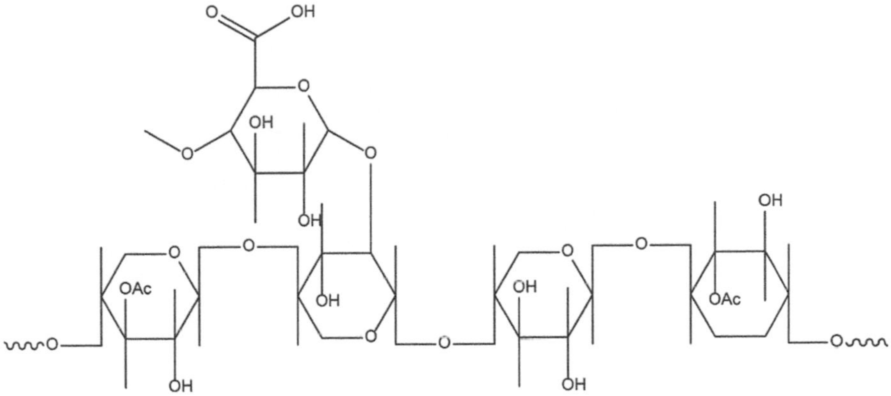

FIGURE 4.12   One of three possible typical hemicellulose structure.

4.4.5.4   LIGNIN

One of the predominant components of wood is lignin which is amorphous in nature. It comprises of asymmetric structure (Figure 4.13) having the most characteristic building blocks as compared to cellulose. That structure variates with source and involves depolymerization mechanism during the isolation process. During the pulping method for paper, making lignin fragments with sulfonate moieties is produced by the delignification mechanism based on the use of sulfites. Due to delignification cellulose, fiber isolates. Dissolved lignin can be utilized as fuel in this context. It enhances the process and also helps the inorganic catalyst to rejuvenate.

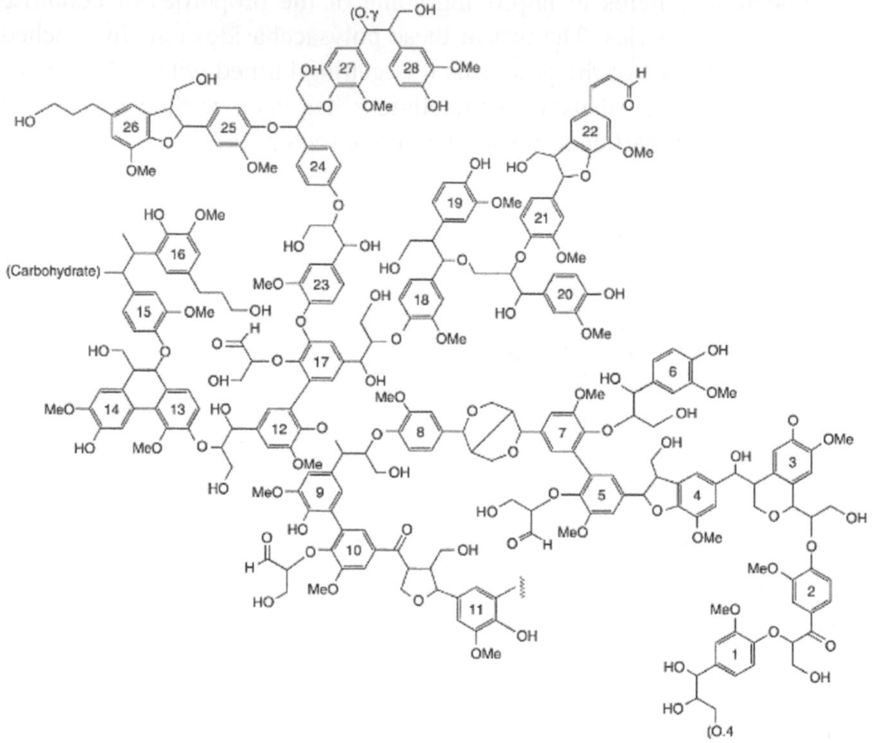

FIGURE 4.13   Polymer assembly of lignin.

Source: Reprinted with permission from Ref. [159]. © 2008 Elsevier.

Considering the mechanical properties, the exiting cellulose fiber has been a paradigm for the existing synthetic fibers [153, 160]. But still, there are existing shortcomings that ensue. There can be incompatibility issues with the polymer matrix due to hydrophobicity of the polymer chain. There is also a tendency of forming aggregates during processing. Biomaterials have a tendency of moisture absorption that can reduce the potential of these materials as reinforcing filler in case of thermoplastic system. Having said that, the design parameter for a polymer composite succumbs to the adhesion property of the fiber with the base polymer material and also the frictional coefficient of the fiber. Shalwan et al. have studied the ways to increase the adhesion between the fiber and the base polymer. According to them, one of the easier ways to achieve that is by treating with a base namely sodium hydroxide. Any other ways which affect directly on the mechanical strength of the fiber, are easily excluded [161]. In order to avoid that Gassan et al.

and Ragoubi et al. have reported strategies based on the physical treatment to improve the polymer/natural fiber compatibility [162, 163]. Li et al. and Belgacem et al. have worked on the surface structure of the fibers and modified them to reduce hydrophobicity and to a certain polarity to achieve higher polar interaction [164, 165].

In terms of electronic devices, cellulose nanofibers (CNF) find its utility. Electrical conductivity and the surface area of the cellulose nanofiber are very high compared to other fibers including some of the synthetic ones. So they find their usage in the field of electrical storage systems and other nano-electronic components [166]. Masoodi et al. has done some work on CNF. They have studied that cellulose finds its usage as a layered film in a bio-resin [167]. For the same purpose sometimes, silica is also deployed along with [168]. Singh et al. [169], De Maura et al. [170], Youssef et al. [171], and Kalia et al. [172] have gone through the cellulose nanocomposites and their utilization as an unconventional packaging material. The characterization of the composites is an essential tool and for that simple polymer characterization techniques such as Fourier transform infrared (FTIR), thermogravimetric analysis (TGA), and differential scanning calorimetry (DSC) are adopted and for the morphological study scanning electron microscope (SEM) has been used. By using a conducting polymer of different concentrations in the composite, one can vary electrical conductivity of the composite and it can have a potential use as an anti-static or anti-bacterial packaging material. Now from a different aspect nanocellulose have two different but important features. Sonic velocity of nanocellulose is high and it arrests the frictional loss. Actually, it has been observed that cellulose transparent film have similar sonic velocity as compared to aluminum and titanium [173]. Jonas et al. [174] have reported that SONY has successfully used it in its headphone diaphragms. These nanocellulose diaphragms are of a thickness of 20 µm and are prepared by dehydration followed by compression. The beauty of nanocellulose is lying in such kind of application only where its similar sound velocity as compared to metallic diaphragm comes into action [176].

When it comes to food and other non-food packaging industry, lignocellulose material finds its application [158]. It can be used for packaging of solid as well as liquid or for medical and pharmaceutical packaging [176, 177]. Nano-cellulose imparts its gas barrier and heat stability properties on polymer matrix. So, researchers have utilized this property of nanocellulose for generating new generation packaging material [178–183]. Reinforcing CNF offer high performance application in case of biofiber based bio-composites [184]. John and Thomas et al. [185] have stated that the amalgamation of nano-cellulose fiber with the renewable polymers produces bio-composites having the vortex as biofiber. These materials are compatible to human body and hence find its utilization in various artificial organs, mostly in kidney dialysis membrane and tubing, blood bags, artificial skin etc. Pineapple leaf fiber and polyurethane composites find its application in various medical implants [186]. Addition of small amount of CNF to polyurethane can increase the strength and stiffness of the product. PVA based composite with bacterial cellulose fibers are also compatible to human body and find its usage in the cardiology. There are reporting of other bionanomaterials for cosmetics, veterinary, dental, drug delivery, medical implants, wound heal dressing, tissue, and cell culture [187–200].

For meeting consumer need as well as regulatory requirements, the highly required field is performance improvement of automobiles. Nanocellulose, thermoplastic composite materials are largely used in automobile industry as it meets the performance needs and car safety. In the case of automotive sector nanocellulose, hemicellulose, lignin fiber can deliver high performance and contribute towards physical and chemical properties at a minimal cost.

Natural fiber draw attention as reinforcing fiber in polymer composites from different thermoplastic matrix such as polypropylene (PP), polyethylene, nylon for other kinds of automotive application. As a further move now, a day's automakers are using these natural fibers composite thermoset matrices for some of their car components [201–205]. According to Shinoj et al. [206], Mercedes Benz is using coconut fiber for seat parts along with NR latex. Some companies find the usage of epoxy resin reinforced with flax or sisal fiber mat for door panels. According to Suddell group, Audi, and Ford use composites from polyurethane matrix reinforced with flax/sisal/kenaf mat for door trim panels [207, 208].

4.4.5.5 CLAY

Clay is a naturally abundant material which has a huge potential in tire industries. Clay minerals are crystalline mostly layered aluminosilicates. It may contain certain amount of iron, alkaline metals, and alkaline earth metals. The crystalline lattice can be derived as mono-, bi-, and tri-dimensional types. Clays are consisting of alternate structure of silicate and alumina. Both this tetrahedral and octahedral structure is bound together by oxygen atom. When the clay is classified as 1:1 clay that means in each layer there is one tetrahedral and one octahedral unit present. When it is 1:2, that indicates one tetra sheet and two octa-sheet in each layer. Similarly, it will change in case of 2: 1 structured clay.

Smectite is in 2:1 phyllosilicate family with a silicate-gibbsite-silicate stacking. Different example of smectite is montmorillonite, saponite, bentonite, vermiculite, laponite, etc. Therefore, hydroxides groups are exposed at one side and the other side by an oxygen atom (Figure 4.14).

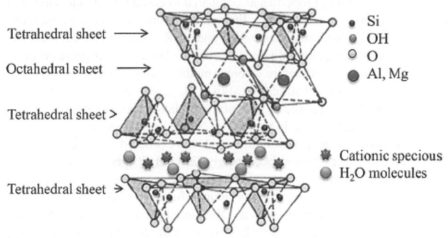

FIGURE 4.14   Crystal stacking structure of Smectite clay (2:1).

Source: Reproduced with permission from Ref. [209]. © 2011. Royal Society of Chemistry.

Kaolinite is 1:1 type clay. A very high cohesion results in kaolinite layers and that hinders intercalation. Another class of clay does exist that is known as layered double hydroxides or anion clay. Brucite is one of the examples of LDH. In brucite magnesium ions are connected with OH groups, building regular octahedron that are linked with one another, forming layers which interact with each other by very week forces. Mostly all clays are hydrophilic that makes them not suitable for mixing. Hence, it is essential to modify the clay before mixing. One of the popular techniques is by ion exchanging. Small cations can replace the big cations. For example, in some cases sodium is being replaced by ammonium ion. Elastomer/rubber is one of the most promising materials for the preparation of PCNs. There are a lot of studies have been done related to PCNs.

Varghese et al. [210] have prepared NR based nanocomposites with 10 wt% bentonite and layered silicate for the latex compounding method. The properties are being compared with the commercial clay. In mechanical, thermal or in swelling test, layered silicate performs much better than commercial clay. Wang et al. [211] have studied NR-MMT and CR-MMT by coagulating latex and clay suspension. The properties are also being compared with NR carbon black composite instead of clay. NR-MMT nanocomposites exhibit higher hardness, higher modulus, high tear strength, excellent gas barrier properties. Higher aspect ratio of clay layers is responsible for diffusing gas and hence have better anti-aging properties as compared to carbon black one. Hence, it finds it application in mostly inner liner, dumper, and inner tube compounds. Sadhu et al. [212] have studied the SBR and NBR system with a combination of copolymer and organomodifier. Unmodified clay did not disperse properly forming agglomerated structure whereas modified clay can disperse properly because of its exfoliated structure. Maji et al. [213] have studied the effect of Cloisite 30B. It is proved to be that 8 phr clay-based nanocomposite can possess two times increase in properties. Several researchers such as Khalil et al.; Ezema et al. and Aleksandra et al. have studied the effect of filler, its content, and size on the rubber matrix [214–216]. Incorporation of nano clay filler into the rubber matrix lead to an improve in mechanical properties. Mechanical properties improve mainly due to the better polymer filler interaction. Rubber interacts with the functional groups present on the clay surface. Rubber clay nanocomposites are basically having an application in the tire tread and base material because rubber clay composites help to gain certain properties like lower weight, lower energy dissemination, and higher air retaining capability. It also helps to extend the magic triangle performance (rolling resistance, traction, and abrasion) of a tire tread [216]. By reviewing several literatures it has been found that clay incorporates longer life into the compound as well as lower fuel consumption without compromising tread grip for winter application [217–219].

Steel cords are used for the radial tire application. So, there is a certain requirement for bringing this reinforcing material under the sustainability umbrella. The question is how can steel cord be sustainable. Steel cord production giant Bekaert have claimed to produce a lower weight steel cord (super tensile or ultra-tensile steel cord) with equivalent strength. That can reduce the tire weight and hence with proper air inflation, tire will have lower rolling resistance which increase fuel efficiency. Lots of work are conducting worldwide related to that. Longer tire life with improve adhesion can definitely lead to a sustainable tire. Beside that to improve adhesion, sometime inorganic cobalt salts are getting mixed in compound stage with the ply compounding formulation. To avoid that TAWI has been introduced where the brass coating contains not only Copper and Zinc but certain amount of Co as well. That certainly a more sustainable approach than earlier [220].

4.5 CONCLUSION

This book chapter is not first of its kind since several articles has been already published regarding this. In the above text, we have set our target pretty simple. We have tried to cover more and gather as much as information possible related to renewable and sustainable source of material which can act as potential replacement for tire industries. Most of the world leading tire industries has taken their goal set for complete replacement of the petroleum-based materials with sustainable eco-friendly materials to have a better future for the next generation to come.

"If it can't be reduced, reused, repaired, rebuilt, refurbished, refinished, resold, recycled, or composted, then it should be restricted or removed from production."

—Pete Seeger (Folk Singer and Social activist)

KEYWORDS

- **biomass**
- **cellulose fiber**
- **dandelion**
- **differential scanning calorimetry**
- **emulsion styrene-butadiene rubber**
- **guayule**
- **liquid Farnesene rubber**
- **multi-functional additives**
- **natural rubber**
- **recycle**
- **sustainability**

REFERENCES

1. Mukhopadhyay, R., Sengupta, R., & Das, Gupta, S., (2008). Recent advances in eco-friendly technology. In: Bhowmick, A. K., (ed.), *Current Topics in Elastomer Research* (p. 1021). CRC Press, Taylor, and Francis Group: NW.
2. ARIES (The Australian Research Institute for Environment and Sustainability). Department of Environmental Science, Macquive University North Ryde, Australia (Accessed on 3 August 2020).
3. https://www.mcgill.ca/sustainability/files/sustainability/what-is-sustainability.pdf (accessed on 10 August 2020).
4. Sutton, P., (2004). *A Perspective on Environmental Sustainability* (pp. 1–32). Paper on the Victorian Commissioner for Environmental Sustainability.
5. Koltun, P., (2010). Materials and sustainable development. *Progress in Natural Science: Materials International, 20,* 16–29.
6. Jagadale, S. C., et al., (2015). Environmental concern of pollution in rubber industry. *International Journal of Research in Engineering and Technology, 4*(11), 187–191.
7. Wagner. I., (2017). *Statistics and Facts on the Global Automotive Industry.*
8. Haraguchi, T., (2006). *Trends in Automobiles and Rubber Parts* (pp. 151–157). KGK.
9. Haraguchi, T., (2006). *Trends in Rubber Parts for Automobiles* (Vol. 79, No. 3, p. 103). Nippon GomuKyokaishi.
10. Haraguchi, T., (2010). The prospects of reducing fuel consumption with automobile basic efficiency. In*: Proceedings of FISITA, Scientific Society for Mechanical Engineering (GTE), F2010-A-013.*
11. Mukhppadhyay, R., (2007). Future perspective of Indian automotive rubber component industry. *Indian and International Rubber Journal.*
12. Aryton, B. A., (2017). *Smart Mobility.* New technology in tires: Electrical tires are the Future.
13. Wills, H., (2014). *The Future of Tires: The Latest Tech and Eco Tires.* A.E.
14. Crawfords, The Future of Tyre Technology., (2016). *In Industry News and Info.*
15. James Burbank the London Economic, The Future of Tyre Technology., (2018). *In Automotive, Tech. and Auto.*
16. David, S., (2014). Key risks for sustainable tyre business., Tire *Asia.*
17. Jones, K. P., (1997). Rubber and Environment., *UNCTD/IRSG Workshop* (pp. 8–19). Manchester.
18. http://content.indiainfoline.com/wc/archives/sect/tyre/ch03.html, India infoline tyre newsletter (2004).
19. Brown, K., https://www.rubbernews.com/suppliers/trinseo-innovation-key-sustainable-tires Rubber and Plastic news (accessed on 3 August 2020).
20. Rasutis, D., et al., (2015). A sustainability review of domestic rubber from the guayule plant. *Industrial Crops and Products, 70,* 383–394.
21. Beilen, J. V. B., & Poirer. Y., (2007). *Crit. Rev. Biotechnol., 27.*
22. https://newatlas.com/russian-dandelion-guayule-latex/23362/ (accessed on 3 August 2020).
23. Santangelo, P., & Roland, C., (2001). The fatigue life of Hevea and guayule rubbers. *Rubber Chemistry and Technology, 74*(1), 69–78.

24. Bhowmick, A., Rampalli, S., & McIntyre, D., (1985). Effect of resin components on the degradation of guayule rubber. *Journal of Applied Polymer Science, 30*(6), 2367–2388.

25. Sarkar, P., & Bhowmick, A. K., (2018). Sustainable rubbers and rubber additives. *Journal of Applied Polymer Science, 135*(24), 45701.

26. Behr, A., & Johnen, L., (2009). Myrcene as a natural base chemical in sustainable chemistry: A critical review. *Chem. Sus. Chem: Chemistry and Sustainability Energy and Materials, 2*(12), 1072–1095.

27. Sarkar, P., & Bhowmick, A. K., (2016). Green approach toward sustainable polymer: Synthesis and characterization of poly(myrcene-co-dibutyl itaconate). *ACS Sustain. Chem. Eng., 4,* 2129–2141.

28. Newmark, R. A., & Majumdar, R. N., (1988). 13C-NMR spectra of cis-polymyrcene and cis-polyfarnesene. *Journal of Polymer Science Part A: Polymer Chemistry, 26*(1), 71–77.

29. Liquid farnesene rubber used in car tires for first time. https://www.kuraray.com/release/2017/170220.html, (accessed on 3 August 2020).

30. Abdelrahman, O. A., et al., (2017). Biomass-derived butadiene by dehydra-decyclization of tetrahydrofuran. *ACS Sustainable Chemistry and Engineering, 5*(5), 3732–3736.

31. Azeem, M., Borg-Karlson, A. K., & Rajarao, G. K., (2013). Sustainable bio-production of styrene from forest waste. *Bioresource Technology, 144,* 684–688.

32. McKenna, R., & Nielsen, D. R., (2011). Styrene biosynthesis from glucose by engineered *E. coli. Metabolic Engineering, 13*(5), 544–554.

33. AMYRIS, Biopharma, hhtp://amyris.com/products.biopharma (Accessed on November 2019).

34. Leavell, M. D., McPhee, D. J., & Paddon, C. J., (2016). Developing fermentative terpenoid production for commercial usage. *Current Opinion in Biotechnology, 37,* 114–119.

35. Ajikumar, P. K., et al., (2008). Terpenoids: Opportunities for biosynthesis of natural product drugs using engineered microorganisms. *Molecular Pharmaceutics, 5*(2), 167–190.

36. Yoo, T., & Henning, S. K., (2017). Synthesis and characterization of farnesene-based polymers. *Rubber Chemistry and Technology, 90*(2), 308–324.

37. Buhl, T., (2014). Global bioenergies announces break-through in direct biological production of butadiene http://www.businesswire.com/news/home/20141126005551/en/Global-Bioenergies-Announces-Break-Through-Direct-Biological-Production (accessed on 3 August 2020).

38. Prochon, M., & Yves, H. T. N., (2017). *Rubber Chem. Technol., 88*(2), 258–275.

39. Barone, J. R., & Schmidt, W. F., (2006). Effect of formic acid exposure on keratin fiber derived from poultry feather biomass. *Bioresource Technology, 97*(2), 233–242.

40. Barone, J. R., & Schmidt, W. F., (2005). Polyethylene reinforced with keratin fibers obtained from chicken feathers. *Composites Science and Technology, 65*(2), 173–181.

41. Barone, J. R., Schmidt, W. F., & Liebner, C. F., (2005). Thermally processed keratin films. *Journal of Applied Polymer Science, 97*(4), 1644–1651.

42. Przepiórkowska, A., Chronska, K., & Prochon, M., (2008). Biodegradable elastomeric vulcanizates. *Rubber Chemistry and Technology, 81*(4), 723–735.

43. Whaley, W. G., & Bowen, J. S., (1947). *Russian Dandelion (Kok-Saghyaz), an Emergency Source of Natural Rubber.* U.S. Department of Agriculture: Washington D.C.

44. Rubber Journal Asia., (2017). *Green Rubber: New Bio Rubber Tires with Greener Footprint.*
45. Burgundy, M., The new tires are bio-based., *The Bio Journal.* http://thebiojournal.com/the-new-tires-are-bio-based/ (accessed on 3 August 2020).
46. *The Smithers Report, September.,* (2019)., *32*(39e).
47. Anthony, A., (2018). *Michelin Researching Eco-Friendly Wood Based Tires.*
48. *Michelin Aims High for Use of Sustainable Materials and Recycling of Tyres.* Tire News A.A-2019. https://tyrenews.com.au/michelin-aims-high-for-use-of-sustainable-materials-and-recycling-of-tyres/ (accessed on 3 August 2020).
49. Hara, Y., & Kirino, Y., (2012). Environmental compound technology for tires. *Nippon Gomu Kyokaishi, 85*(6), 187–192.
50. Simon, S., (2019). https://www.cyclingweekly.com/news/product-news/continental-launches-bike-tyre-made-dandelions-424634 (accessed on 3 August 2020).
51. Rich, A., (2016). *Eco Friendly Tires; Tire Review Magazine.* https://www.tirereview.com/157213/ (accessed on 3 August 2020).
52. Crane, Elefritz, R. A., Kay, E. L., & Laman. J. R., (1978). *Rubber Chem. Technol., 51*(3), 577–599.
53. https://www.waste360.com/recycling/look-what-s-ahead-scrap-tire-recycling (accessed on 3 August 2020).
54. http://www.statista.com/statistics/614554/scrap-tire-generation-share-by-type-in-the-united states/ (Accessed on November 2019).
55. https://waste-management-world.com/a/scrap-tyre-recycling (accessed on 3 August 2020).
56. Beckman, J., et al., (1974). Scrap tire disposal. *Rubber Chemistry and Technology, 47*(3), 597–624.
57. Brown, D. A., Watson. W. A., (2000). *Novel Concepts in Environmentally Friendly Rubber Recycling.* Presented at the International Rubber Forum 2000, Antwerp.
58. Srinivasean, A. Shanmugharaj, A. M., & Bhowmick, A. K., (2008). Waste rubber recycling. In: Bhowmick, A. K., (ed.), *Current Topics in Elastomer Research* (p. 1043). Taylor and Francis Group, Boca Rayton, FL.
59. Bredberg, K. P. J., Christiansson, M., Stenberg, B., & Holst, O., (2001). *Appl. Microbiol. Biotechnol., 55,* 43.
60. Aoudia, K., et al., (2017). Recycling of waste tire rubber: Microwave devulcanization and incorporation in a thermo set resin. *Waste Management, 60,* 471–481.
61. Jana, G. K., & Das, C. K., (2005). *Polymer Plast. Technol. Eng.,* 44.
62. Jana, G. K., & Das, C. K., (2005). *Prog. Rubber Plast. Recycling Technol.,* 21.
63. Rajan, V., et al., (2005). Comparative investigation on the reclamation of NR based latex products with amines and disulfides. *Rubber Chemistry and Technology, 78*(5), 855–867.
64. Rajan, V., et al., (2005). Model compound studies on the devulcanization of natural rubber using 2,3-dimethyl-2-butene. *Rubber Chemistry and Technology, 78*(4), 572–587.
65. Rajan, V. V., & Dierkes, W. K., & Noordermeer, J. W. M., (2006). *J. Appl. Polym. Sci.,* 102.
66. Kojima, M., Tosaka, M., & Ikeda, Y., (2004). Chemical recycling of sulfur-cured natural rubber using supercritical carbon dioxide. *Green Chemistry, 6*(2), 84–89.
67. De, D., et al., (2006). Reclaiming of ground rubber tire (GRT) by a novel reclaiming agent. *European Polymer Journal, 42*(4), 917–927.

68. De, D., & De, D., (2011). Processing and material characteristics of a reclaimed ground rubber tire reinforced styrene butadiene rubber. *Materials Sciences and Applications, 2*(05), 486.

69. De, D., De, D., & Singharoy, G., (2007). Reclaiming of ground rubber tire by a novel reclaiming agent. I. Virgin natural rubber/reclaimed GRT vulcanizates. *Polymer Engineering and Science, 47*(7), 1091–1100.

70. Verbruggen, M. A. L. Devulcanization of EPDM Rubber, (2007). PhD Thesis. University of Twente, Enschede, The Netherlands.

71. Yehia, A. A. I., M. N., Hefny, Y. A., Abdelbary, E. M., & Mull, M. A., (2004*). J. Elast. Plast.*, 33.

72. Meyers, R. D., Nicholson, P., MacLeod, J. B., & Moir, M. E., (1997). *U.S. Patent 5,602,186.* (To Exxon Research and Engineering Company).

73. Grigoryeva, O., et al., (2008). Reactive compatibilization of recycled polyethylenes and scrap rubber in thermoplastic elastomers: Chemical and radiation-chemical approach. *Rubber Chemistry and Technology, 81*(5), 737–752.

74. Amash, A., Giese, U., & Schuster. R., (2002). Interphase grafting of reclaimed rubber powder. *Kautschuk Gummi Kunststoffe, 55*(5), 218–218.

75. Rezaei, A. M., Jalali, A. A., & Nazockdast, H., (2010). Partial replacement of NR by GTR in thermoplastic elastomer based on LLDPE/NR through using reactive blending: Its effects on morphology, rheological, and mechanical properties. *Journal of Applied Polymer Science, 115*(4), 2416–2422.

76. Bredberg, K., et al., (2001). Anaerobic desulfurization of ground rubber with the thermophilic archaeon *Pyrococcus furiosus*-a new method for rubber recycling. *Applied Microbiology and Biotechnology, 55*(1), 43–48.

77. Romine, R. A., Romine, M. F., & Snowden-Swan, L., (1995). *Microbial Processing of Waste Tire Rubber.* Paper 56, presented at the 148th fall meeting of the rubber division, ACS, Cleveland.

78. http://tirebusiness.com/article/19950626/ISSUE/306269960/researchers-find-bacteria-that-eat-tiresfor-lunch, (1995). *Tire-Eating Bacteria Under Study.*

79. Beckman, J. A. C., Kay, E. L., & Laman, J. R., (1972). Akron rubber group meeting. *Rubber Age, 105*(4), p. 43.

80. IeBeau, D. S., (1967). *Rubber Chem. Technology.*, (Vol. 40, p. 217).

81. Manuel, H. J., & Dierkes, W., (1995). *RAPRA Review Report.* IRSG.

82. Glenn, R. C., Ward, C. Y., (1972). *Presented at the Southern Agricultural Workers Conference.* Richmond, Virginia.

83. Mandal, N., Gupta, S. D., & Mukhopadhyaya, R., (2005). Regeneration of carbon black from waste automobile tires and its use in carcass compound. *Progress in Rubber Plastics and Recycling Technology, 21*(1), 55–72.

84. Bandyopadhyay, S., et al., (2006). Use of recycled tire material in NR/BR blend based tire tread compound: Part II (with ground crumb rubber). *Progress in Rubber, Plastics, and Recycling Technology, 22*(4), 269–284.

85. Bandyopadhyay, S., et al., (2006). Effect of regenerated carbon-black on a bias tire tread cap compound. *Progress in Rubber, Plastics, and Recycling Technology, 22*(3), 195–210.

86. Dasgupta, S., Agrawal, S., Bandyopadhyay, S., Chakraborty, S., Mukhopadhyay, R., Malkani, R. K., & Ameta, S. C., (2007). *Polymer Testing, 26,* 489.

87. Dasgupta, S., et al., (2008). Characterization of eco-friendly processing aids for rubber compound: Part II. *Polymer Testing, 27*(3), 277–283.

88. Dasgupta, S., et al., (2009). Eco-friendly processing oils: A new tool to achieve the improved mileage in tire tread. *Polymer Testing, 28*(3), 251–263.

89. Dasgupta, S., et al., (2009). Improved polymer-filler interaction with an eco-friendly processing aid: Part 1. *Progress in Rubber Plastics and Recycling Technology, 25*(3), 141–164.

90. *Encyclopedia of Polymer Science and Engineering, Cellular Materials to Composites, Second Edition, A Wiley-Interscience Publication,* (1985), vol 3, 619.

91. Joseph, R., Madhusoodhanan, K. N., Alex, R., Varghese, S., George, K. E., & Kuriakose, B., (2004). *Plastics Rubbers Compos., 33,* 217.

92. Nag, A., Halder, S. K., (2006). *Kautschuk Gummi Kunstsoffe,* 322.

93. Botros, S. H., El-Mohsen, F. F., & Meinecke, E. A., (1987). *Rubber Chem. Technol., 60,* 159.

94. Ismail, H., (2000). Effect of palm oil fatty acid additive (POFA) on curing characteristics and vulcanizate properties of silica filled natural rubber compounds. *Journal of Elastomers and Plastics, 32*(1), 33–45.

95. Wang, Z., et al., (2016). Investigation of palm oil as green plasticizer on the processing and mechanical properties of ethylene propylene diene monomer rubber. *Industrial and Engineering Chemistry Research, 55*(10), 2784–2789.

96. Pechurai, W., Chiangta, W., & Tharuen, P., (2015). Effect of vegetable oils as processing aids in SBR compounds. In: *Macromolecular Symposia.* Wiley Online Library.

97. Petrović, Z. S., et al., (2013). Soybean oil plasticizers as replacement of petroleum oil in rubber. *Rubber Chemistry and Technology, 86*(2), 233–249.

98. Sahakaro, K., & Beraheng, A., (2011). Epoxidized natural oils as the alternative safe process oils in rubber compounds. *Rubber Chemistry and Technology, 84*(2), 200–214.

99. Li, J., et al., (2018). Sustainable plasticizer for butyl rubber cured by phenolic resin. *Journal of Applied Polymer Science, 135*(24), (4550). 0.

100. Fernandez, S. S., Kunchandy, S., & Ghosh, S., (2015). Linseed oil plasticizer based natural rubber/expandable graphite vulcanizates: Synthesis and characterizations. *Journal of Polymers and the Environment, 23*(4), 526–533.

101. Nanoshel, (2019). *Application of Zinc Oxide for Industrial Rubber and Plastics.* Zinc oxide for Rubber Industries.

102. Walter, J., (2009). *Tire Technology International,* 18.

103. Chapman, A. V., & Johnson, T., (2005). *Kautschuk Gummi Kunststoffe, 58,* 358.

104. Sahoo, S. & Bhowmick, A. K., (2007). *J. Appl. Polym. Sci., 106,* (3077).

105. Przybyszewska, M., Zaborski, M., Jakubowski, B., & Zawadiak, J., (2009). *Express Polym. Lett., 3,* 256.

106. Guzmán, M., et al., (2012). Zinc oxide versus magnesium oxide revisited: Part 2. *Rubber Chemistry and Technology, 85*(1), 56–67.

107. Bhatia, K., & Ayyar, K. S., (1980). *Indian Forester, 106,* 363.

108. Bhatia, K. L., & Swaleh. M., (1980). *Indian Forester, 103,* 273.

109. Bhatia, K. L., & Swaleh, M., (1980). *Indian Forester, 107,* 519.

110. Guhathakurta, S., et al., (2006). Waste natural gum as a multifunctional additive in rubber. *Journal of Applied Polymer Science, 102*(5), 4897–4907.

111. Saha, S., et al., (2005). Studies of *Terminalia bellerica* (Bahera), a natural gum, as an additive in a water-based adhesive composition. *Journal of Adhesion Science and Technology, 19*(15), 1349–1361.
112. Shanmugharaj, A., et al., (2005). Studies on the adhesion behavior of water-based adhesives blended with as a gum. *Journal of Adhesion Science and Technology, 19*(8), 639–658.
113. Hurlston, E., (1936). The use of factice in rubber manufacture. *Rubber Chemistry and Technology, 9*(4), 621–625.
114. Nandanan, V., Joseph, R., & George, K., (1999). Rubber seed oil: A multipurpose additive in NR and SBR compounds. *Journal of Applied Polymer Science, 72*(4), 487–492.
115. Bhattacharyya, S. K., et al., (2012). Exploring microcrystalline cellulose (mcc) as a green multifunctional additive (MFA) in a typical solution-grade styrene butadiene rubber (S-SBR)-based tread compound. *Industrial and Engineering Chemistry Research, 51*(32), 10649–10658.
116. Manoharan, P., & Naskar, K., (2017). Biologically sustainable rubber resin and rubber-filler promoter: A precursor study. *Polymers for Advanced Technologies, 28*(12), 1642–1653.
117. Pinela, J., et al., (2017). Valorization of tomato wastes for development of nutrient-rich antioxidant ingredients: A sustainable approach towards the needs of the today's society. *Innovative Food Science and Emerging Technologies, 41*, 160–171.
118. Alasalvar, C., & Shahidi, F., (2009). Natural antioxidants in tree nuts. *European Journal of Lipid Science and Technology, 111*(11), 1056–1062.
119. Wu, S., Weng, P., Tang, Z., & Guo, B., (2016). *ACS Sustainable Chem. Eng., 4*, 247.
120. Zhan, K., Ejima, H., & Yoshie, N., (2016). Antioxidant and adsorption properties of bioinspired phenolic polymers: A comparative study of catechol and gallol. *ACS Sustainable Chemistry and Engineering, 4*(7), 3857–3863.
121. Tran, T. N., et al., (2017). Starch-based bio-elastomers functionalized with red beetroot natural antioxidant. *Food Chemistry, 216*, 324–333.
122. Lin, T., et al., (2015). Renewable conjugated acids as curatives for high-performance rubber/silica composites. *Green Chemistry, 17*(6), 3301–3305.
123. Ding, C., & Matharu, A. S., (2014). Recent developments on bio-based curing agents: A review of their preparation and use. *ACS Sustainable Chemistry and Engineering, 2*(10), 2217–2236.
124. Gomez, I., et al., (2016). Inverse vulcanization of sulfur using natural dienes as sustainable materials for lithium-sulfur batteries. *Chem. Sus. Chem., 9*(24), 3419–3425.
125. Ten, B. A., (2002). *Silica Reinforced Tire Rubbers: Mechanistic Aspects of the Role of Coupling Agents*. University of Twente: Enschede.
126. Kraus, G., (1978). Reinforcement of elastomers by particulate fillers. In: Eirich, F. (ed.), *Science*, and *Technology of Rubber*. Academic Press: New York.
127. Wischhusen, M., (2010). *Balancing Rolling Resistance with Other Performance Characteristics*. Tire Review.
128. *Organization for Economic Cooperation and Development (OECD)*, (2004). SIDS Initial assessment Report for SIAM 19: Synthetic Amorphous Silica and Silicates.
129. Schlomach, J., & Kind, M., (2004). Investigations on the semi-batch precipitation of silica. *Journal of Colloid and Interface Science, 277*(2), 316–326.

130. Integrated Pollution Prevention and Control, (2006). Reference document on best available techniques for the manufacture of large volume inorganic chemicals-solid and other industry, European Commission.

131. https://www.pneurama.com/en/rivista_articolo.php/Nanomaterials-a-tire-s-biggest-friends?ID=36715 (accessed on 3 August 2020).

132. Michelin Innovation. http://www.michelin.in/IN/en/whymichelin/more-on-brand-innovation.html, (accessed on 3 August 2020).

133. Tire Solutions from Solvay. http://www.solvay.com/en/binaries/TIRE % 20SOLUTIONs-177294.pdf (accessed on 3 August 2020).

134. Highly dispersible silica from EVONIK. http://ultrasil.evonik.com/product/ultrasil/en/products/highly dispersiblesilica/pages/default.aspx, (accessed on 3 August 2020).

135. PPG Silica Products. http://www.ppgsilica.com/getmedia/0de26f2f-8399-492e-9170-5a492d38c6c5/HiSiEZ200G.pdf.aspx (accessed on 3 August 2020).

136. Wang, M. J., et al., (2002). New generation carbon-silica dual phase filler part I. Characterization and application to passenger tire. *Rubber Chemistry and Technology, 75*(2), 247–263.

137. Nayek, S., et al., (2005). Wear behavior of silica filled tire tread compounds by various rock surfaces. *Rubber Chemistry and Technology, 78*(4), 705–723.

138. Lodha, V., Bhattacharyya, S., Dasgupta, S., Mukhopadhyay, R., Sarkar, P., & Bhowmick, A. K., (2017). In: *The Proceedings of the 192nd Technical Meeting of the Rubber Division.* Cleveland, OH: ACS.

139. Asim, M., Abdan, Khalina., Jawaid, M., et al., (2015). A review on pineapple leaves fiber and its composites. *nt. J. Polym. Sci.,* 16.

140. Jawaid, M., & Abdul, Khalil, H. P. S., (2011). Cellulosic/synthetic fiber reinforced polymer hybrid composites: A review. *Carbohydrate Polymers, 86*(1), 1–18.

141. Lamaming, J., et al., (2015). Isolation and characterization of cellulose nanocrystals from parenchyma and vascular bundle of oil palm trunk (*Elaeisguineensis*). *CarbohydrPolym., 134*, 534–540.

142. Morton, W. E., & Hearle, J. W. S., (1962). *Physical Properties of Textile Fibers.* Butterworth & Co. Ltd: London.

143. Abdul, K. H. P. S., Bhat, A. H., & Ireana, Y. A. F., (2012). Green composites from sustainable cellulose nanofibrils: A review. *Carbohydrate Polymers, 87*(2), 963–979.

144. Ishikawa, A., Okano. T., & Sugiyama, J., (1997). Fine structure and tensile properties of ramie fibers in the crystalline form of cellulose I, II, III, and IV. *Polymer, 38*(2), 463–468.

145. Vanderhart, D. L., & Atalla, R. H., (1984). Studies of microstructure in native celluloses using solid-state carbon-13 NMR. *Macromolecules, 17*(8), 1465–1472.

146. Isogai, A., (2013). Wood nanocelluloses: Fundamentals and applications as new bio-based nanomaterials. *Journal of Wood Science, 59*(6), 449–459.

147. Gray, D. G., (1974). Polypropylene transcrystallization at the surface of cellulose fibers. *J. Polym. Sci. Polym. Lett. Edn., 12*(9), 509–515.

148. McAllister, D. H., Pearson, P., & Wells, H., (1982). *Proceedings of the Reinforced Plastics Congress.* Brighton, UK: British Plastics Federation.

149. Abdul Khalil, H. P. S., et al., (2017). 1-Nanofibrillated cellulose reinforcement in thermo set polymer composites. In: Jawaid, M., Boufi, S., & Abdul Khalil, H. P. S., (eds.), *Cellulose-Reinforced Nanofiber Composites* (pp. 1–24) Woodhead Publishing.

150. Bledzki, A. K., & Gassan, J., (1999). Composites reinforced with cellulose based fibers. *Progress in Polymer Science, 24*(2), 221–274.

151. Ansell, M. P. et al., (2001). Review of current international research into cellulosic fibers and composites. *Journal of Materials Science, 36*, 2107–2131.

152. Saheb, D. N., & J. N., (1999). Natural fiber reinforced composites is an emerging area in polymer science. *Adv. in Polymer Techn., 18*, 351–363.

153. Wambua, P. I., J., & Verpoest, I., (2003). Natural fibers: Can they replace glass in fiber reinforced plastics. *Compos. Sci. Technol., 63*(9), 1259–1264.

154. Anandjiwala, R. D., & Blouw. S., (2004). Bast fibrous plants for healthy life. In: *Proceedings of the FAO Global Workshop.* Banja Luka, Bosnia-Herzegovina.

155. Bodros, E., Pillin, I., Montrelay, N., & Baley, C., (2007). Could biopolymers reinforced by randomly scattered flax fiber be used in structural applications? *Compos. Sci. Technol., 67*, 462–470.

156. Carus, M., G. C., Pendarovski, C., Vogt, D., Ortmann, S., Grotenhermen, F., Breuer, T., & Schmidt, C., (2008). Studie zur markt-und konkurrenzsituation bei naturfasern und naturfaserwerkstoffen (Deutschland Und Eu). In: *Fachagentur Nachwachsende Rohstoffe (FNR).* Guelzow.

157. Karus, M., & K. M., (2002). Natural fibers in the European automotive industry. *Article in Journal of Industrial Hemp, 7*(1), 119–131.

158. Dungani, R., et al., (2017). 3-Bionanomaterial from agricultural waste and its application. In: Jawaid, M., Boufi, S., & Abdul Khalil, H. P. S., (eds.), *Cellulose-Reinforced Nanofiber Composites* (pp. 45–88). Woodhead Publishing.

159. Gandini, A., & Belgacem, M. N., (2008). State of the art. In: Gandini, A. M. N. B., (ed.), *Monomers, Polymers, and Composites from Renewable Resources.* Elsevier: Oxford OX5 1GB, UK.

160. Faruk, O., Bledzki, A. K., Fink, H. P., & Sain, M., (2012). 2000–2010, Biocomposites reinforced with natural fibers. *Prog. Polym. Sci., 37*(11), 1552–1596.

161. Shalwan, A., & Yousif. B. F., (2013). In state of art: Mechanical and tribological behavior of polymeric composites based on natural fibers. *Mater. Des., 48*, 14–24.

162. Gassan, J., & G. V., (2000). Effects of corona discharge and UV treatment on the properties of jute-fiber expoxy composites. *Compos. Sci. Technol., 60*(15), 2857–2863.

163. Ragoubi, M., Bicnaime, D., Molina, S., George, B., & Merlin, A., (2010). Impact of corona treated hemp fibers onto mechanical properties of polypropylene composites made thereof *Ind. Crops Prod., 31*(2), 344–349.

164. Li, Z., Wang. L., & Wang, X., (2007). Cement composites reinforced with surface modified coir fibers. J. Compos. Mater., 41(12), 1445–1457. Li, Z., Wang. L., & Wang, X., (2007). Cement composites reinforced with surface modified coir fibers. *J. Compos. Mater., 41*(12), 1445–1457.

165. Belgacem, M. N., & Gandini. A., (2005). The surface modification of cellulose fibers for use as reinforcing elements in composite materials. *Composite Interfaces, 12*(1/2), 41–75.

166. Liu, H., Bai, J., Wang, Q., Li, C., Wang, S., Sun, W., & Huang, Y., (2014). Preparation and characterization of silver nanoparticles/carbon nanofibers via electrospinning with research on their catalytic properties. *Nano Brief Rep. Rev., 9*(3), 145004-1-145004-7.

167. Masoodi, R., El-Hajjar. RF., Pillai, K. M., & Sabo, R., (2012). Mechanical characterization of cellulose nanofiber and bio-based epoxy composite. *Mater. Des., 36*, 570–576.

168. Tabatabaei, S., Shukohfar, A., Aghababazadeh, R., & Mirhabibi, A., (2006). Experimental study of the synthesis and characterization of silica nanoparticles via the sol-gel method. *J. Phys. Conf. Ser., 26,* 371–374.

169. Singh, A., Sharma. P. K., & Malviya, R., (2011). Eco friendly pharmaceutical packaging material. *World Appl. Sci. J., 14*(11), 1703–1716.

170. De Moura, M. R., Mattoso, L. H. C., & Zucolotto, V., (2012). Development of cellulose-based bactericidal nanocomposites containing silver nanoparticles and their use as active food packaging. *J. Food Eng., 109,* 520–524.

171. Youssef, A. M., El-Samahy, M. A., & Abdel, R. M. H., (2012). Preparation of conductive paper composites based on natural cellulosic fibers for packaging applications. *Carbohydrate Polymers, 89*(4), 1027–1032.

172. Kalia, S., et al., (2014). Nanofibrillated cellulose: Surface modification and potential applications. *Colloid and Polymer Science, 292,* 5–31.

173. Iguchi, M., Yamanaka. S., & Budhiono, A., (2000). Bacterial cellulose-a masterpiece of nature's arts. *J Mater Sci., 35*(2), 261–270.

174. Jonas, R., F. L., (1998). Production and application of microbial cellulose. *Polym. Degrad. Stab., 59*(3), 101–106.

175. Kalia, S., Dufresne, A., Cherian, B. M., Kaith, B., Avérous, L., & Njuguna, J., (2011). Cellulose-based bio-and nanocomposites: A review. *Int. J. Polym. Sci.,* 1–35.

176. Nafchi, A. M., et al., (2012). Antimicrobial, rheological, and physicochemical properties of sago starch films filled with nanorod-rich zinc oxide. *Journal of Food Engineering, 113*(4), 511–519.

177. Zadbuke, N., et al., (2013). Recent trends and future of pharmaceutical packaging technology. *Journal of Pharmacy and Bioallied Sciences, 5*(2), 98–110.

178. Lewis, H., Verghese, K., & Fitzpatrick, L., (2010). Evaluating the sustainability impacts of packaging: The plastic carry bag dilemma. *Pack. Technol. Sci., 23,* 145–160.

179. Freire, M. G., Teles, A. R. R., Ferreira, R. A. S., Carlos, L. D., Lopes-Da-Silva, J. A., & Coutinho, J. A. P., (2011). Electrospun nanosized cellulose fibers using ionic liquids at room temperature. *Green Chem., 13,* 3173–3180..

180. Zhang, X., Tu, M., & Paice, M., (2011). Routes to potential bioproducts from lignocellulosic biomass lignin and hemicelluloses. *Bio. Energy Res., 4*(4), 246–257.

181. Ul-Islam M., Khan, T., & Park, J. K., (2012). Nanoreinforced bacterial cellulose-montmorillonite composites for biomedical applications. *Carbohydr. Polym., 89*(4), 1189–1197.

182. Azeredo, H. M. C., (2013). Antimicrobial nanostructures in food packaging trends. *Food Sci. Technol., 30*(1), 56–69.

183. Sirviö, J. A., et al., (2013). Sustainable packaging materials based on wood cellulose. *RSC Advances, 3*(37), 16590–16596.

184. Chakraborty, A., Sain. M., & Kortschot, M., (2006). Cellulose microfibers as reinforcing agents for structural materials; cellulose nanocomposites: Processing, characterization, and properties. In: *ACS Symposium Series.* Washington, DC: ACS.

185. John, M. J., & Thomas. S., (2008). Biofibers and biocomposites. *Carbohydr. Polym., 71*(3), 343–364.

186. Cherian, B. M., Leao, A. L., de Souza. S. F., Thomas, S., Pothan, L. A., & Kottaisamy, M., (2010). Isolation of nanocellulose from pineapple leaf fibers by steam explosion. *Carbohydr. Polym., 81*(3), 720–725.

187. Millon, E., & Wan, W. K., (2006). The polyvinyl alcohol-bacterial cellulose system as a new nanocomposite for biomedical applications. J. Biomed. Mater. Res. B Appl. Biomater., 79(2), 245–253.

188. Costa, L. M. M., de Olyveira, GM., Cherian, B. M., Leão, A. L., de Souza, S. F., & Ferreira, M., (2013). Bionanocomposites from electrospun PVA/pineapple nanofibers/ Stryphnodendron adstringens bark extract for medical applications. Ind. Crops Prod., 41, 198–202.

189. Baumann, M. D., Kang. CE., Stanwick, J. C., Wang, Y., Kim, H., Lapitsky, Y., et al., (2009). An injectable drug delivery platform for sustained combination therapy. J. Control Release, 138(3), 205–213.

190. Nadagouda, M. N., & Varma, R. S., (2008). Green synthesis of silver and palladium nanoparticles at room temperature using coffee and tea extract. Green Chemistry, 10(8), 859–862.

191. Czaja, W. K., et al., (2007). The future prospects of microbial cellulose in biomedical applications. Biomacromolecules, 8(1), 1–12.

192. Iamaguti, L. S., Brandao. CVS., Minto, B. W., Mamprim, M. J., Ranzani, J. J. T., & Gomes, D. C., (2008). Utilização de membrana biossintética de celulose na trocleoplastia experimental emcães. Avaliações clínica, radiográfica e macroscópica. Vet e Zootec, 15(1), 160–218.

193. Macedo, N. L., Matuda. F., Macedo, L. G. S., Monteiro, A. S. F., Valera, M. C., & Carvalho, Y. R., (2004). Evaluation of two membranes in guided bone tissue regeneration: Histological study in rabbits. Braz J. Oral Sci., 3, 395–400.

194. Klemm, D., Schumann. D., Udhardt, U., & Marsch, S., (2001). Bacterial synthesized cellulose- artificial blood vessels for microsurgery. Prog. Polym. Sci., 26, 1561–1603.

195. Ikada, Y., (2006). Challenges in tissue engineering. J. R. Soc. Interface, 3(10), 589–601.

196. Kumar, R., Roopan, S. M., Prabhakarn, A., Khanna, V. G., & Chakroborty, S., (2012). Agricultural waste Annona squamosa peel extract: Biosynthesis of silver nanoparticles. Spectrochim Acta Part A., 90, 173–176.

197. Seal, B. L., Otero, T., & Panitch, A., (2001). Polymeric biomaterials for tissue and organ regeneration. Mater. Sci. Eng. Rep., 34, 147–230.

198. Capes, J. S., Ando, H. Y., & Cameron, R. E. J., (2005). Fabrication of polymeric scaffolds with a controlled distribution of pores. Mater. Sci. Mater. Med., 16, 1069–1075.

199. Bhattacharya, M., Malinen, M., Lauren, P., Lou, Y. R., Kuisma, S. W., Kanninen, L., et al., (2012). Nanofibrillar cellulose hydrogel promotes three-dimensional liver cell culture. J. Control Release, 164(3), 291–298.

200. Hua, K., et al., (2014). Translational study between structure and biological response of nanocellulose from wood and green algae. RSC Advances, 4(6), 2892–2903.

201. Bartus, S. D., Vaidya, U., & Ulven, C. A., (2005). Design and development of a long fiber thermoplastic bus seat. Compos, Struct., 67, 263–277.

202. Jeyanthi, S., & Rani, J. J., (2012). Improving mechanical properties by kenaf natural long fiber reinforced composite for automotive structures. J. Appl. Sci. Eng., 15(3), 275–280.

203. Cicero, J. A., Dorgan, J., Dec, S. F., & Knauss, D. M., (2002). Phosphite stabilization effects on two step melt-spun fibers of polyactide. Polym. Degrad. Stab., 78, 95–105.

204. Mohanty, A. K., Misra, M., & Hinrichsen, B. G., (2000). Biodegradable polymers and biocomposites: An overview. Macromol. Mater. Eng., 276, 277(1), 1–24.

205. Brouwer, W. D., (2000). Natural fiber composites: Where flax compete with glass? *Sampe J., 36*(6), 18–23.
206. Shinoj, S., Visvanathan, R., Panigrahi, S., & Kochubabu, M., (2011). Oil palm fiber (OPF) and its composites: A review. *Ind. Crops Prod., 33*(1), 7–22.
207. Suddell, B. C., & Evans, W. J., (2005). Natural fiber composites in automotive applications in natural fibers in biopolymers and their biocomposites. In: *Natural Fibers, Biopolymers, and Biocomposites*. Boca Raton, FL: CRC Press.
208. Pickering, K., (2008). *Properties and Performance of Natural-Fiber Composites*. Cambridge: Woodhead Publishing.
209. Ghadiri, M., Chrzanowski, W., & Rohanizadeh, R., (2015). Biomedical applications of cationic clay minerals. *RSC Advances, 5*(37), 29467–29481.
210. Varghese, S., & Karger-Kocsis, J., (2003). *Polymer, 44*.
211. Wang, Y., et al., (2005). *J. Appl. Polym. Sci, 96*.
212. Sadhu, S., & Bhowmick, A. K., (2004). Preparation and properties of nanocomposites based on acrylonitrile-butadiene rubber, styrene-butadiene rubber, and polybutadiene rubber. *Journal of Polymer Science Part B: Polymer Physics, 42*(9), 1573–1585.
213. Maji, P. K., Guchhait, P. K., & Bhowmick, A. K., (2008). Effect of the microstructure of a hyperbranched polymer and nanoclay loading on the morphology and properties of novel polyurethane nanocomposites. *ACS Applied Materials and Interfaces, 1*(2), 289–300.
214. Khalil, A., Saikh, N. S., Nudrat, R. Z., & Khalid, M., (2013). Effect of micro-sized marble sludge on physical properties of natural rubber composites. *Chem. Indust. Chem. Eng. Quart., 19*(2), 281–293.
215. Ezema, I. C., Menon, A. R R., Obayi, C. S., & Omah, A. D., (2014). Effect of surface treatment and fiber orientation on the tensile and morpholoical properties of banana stem fiber reinforced natural rubber composite. *JMMCE, 2*(3), 216–222.
216. Aleksandra, I. D., Gordana, B. G., Aleksandra, B., Igor, G., & Luljeta, R., (2014). Preparation and properties of natural rubber/organo-montmorillonite: From lab samples to bulk material. *Macedonian J. Chem. Chem. Eng., 33*(2), 249–265.
217. Arroyo, M., López-Manchado, M. A., & Herrero, B., (2003). Organo-montmorillonite as substitute of carbon black in natural rubber compounds. *Polymer, 44*(8), 2447–2453.
218. Fidelis, C., Piwai, S., Benias, N. C., Upenyu, G., & Mambo, M., (2013). Maize stalk as reinforcement in natural rubber composites. *Int. J. Sci. Technol. Research, 2*(6), 263–271.
219. Ku, H., et al., (2011). A review on the tensile properties of natural fiber-reinforced polymer composites. *Composites Part B: Engineering, 42*, 856–873.
220. https://www.bekaert.com/en/products/automotive/corner/steel-cord-for-tire-reinforcement (accessed on 3 August 2020).

CHAPTER 5

Silk: An Explorable Biopolymer in the Biomedical Arena

ATHIRA JOHN,[1] ANANTHU PRASAD,[2] and NEHA KANWAR RAWAT[3]

[1]Center for Biopolymer Science and Technology (CBPST), CIPET, Kochi, Kerala, India

[2]International and Inter-University Center for Nanoscience and Nanotechnology (IIUCNN), Mahatma Gandhi University, Kottayam, Kerala, India

[3]CSIR – National Aerospace Laboratories Bangalore, Karnataka, India

ABSTRACT

The biopolymer silk as a recognizable implement in the biomedical field is manifested in nanotechnology due to its promising avenues. Even though silk sutures were known for centuries, yet an expansion of this research had taken place in these recent decades. The excellent biocompatibility and slow degradation make it well applicable in this field. Silk is mainly composed of two types of proteins namely fibroin and sericin, each of which have created its benchmark in the biomedical world. Wound dressing, drug delivery, implants, and tissue engineering (TE) is the major fields of applications. Silk blended with various other synthetic and natural polymers have found unique applications in these fields. The prominent techniques like electrospinning have found wide applications in incorporating silk in biomedical field. The chapter discusses the scope for this unexplorable material: silk, its prospects and limitations are also explored.

5.1 INTRODUCTION

The *Bombyx mori* silkworm silk has found its application in textile industries for centuries. However, the biomedical field remained unexplored
since its conception. Fortunately, researchers found the unique properties
of silk to be applicable in biomedical grounds in recent decades and are
still research is under progress [1]. Silk is a novel class of proteins having
high molecular weight. They possess unique features like self-assembly,
robust mechanical properties, biocompatibility, and biodegradability
which makes them more suitable for biomaterials application [2]. It is in
1969, that the first patent available in literature by Bloch and Messores,
proposed the use of silk reconstituted by LiSCN/LiBr as are placement
for the standard wax coating used on silk sutures to reduce their limpness,
fraying, and unwanted capillary action [3]. After this cited, there was mass
research initiated to enhance the properties of silk and its application in the
biomedical arena.

5.2 STRUCTURE AND CONSTITUENTS

Biologically silk can be defined as a structural protein that has been spun
into as fiber for using outside the body. The most well-studied silk is the
one made of the domesticated silkworm (*Bombyx mori*) for the reason that
it possesses a semi-crystalline silk core (silk fibroin: SF) which excellent
load-bearing capacity and an external layer called sericin which functions
as a gumming agent [4]. Typically, silk produced by *Bombyx Mori*, silkworm consists of 70–80% fibroin, 20–30% sericin, 0.2–0.4% wax, and
1.2–1.6% carbohydrates. 0.7% inorganic matter and. 0.2% pigments [6]
(Figure 5.1).

5.2.1 SILK FIBROIN (SF)

Silkworm fibroin is used as biomedical sutures for ages; it is also used in
textile production. It is thin, long, light, and soft. However, one of the most
potent natural fibers. This was due to the chemical structure of the protein
[7]. Fibroin consists of light (L) chain polypeptide, and heavy (H) chain
polypeptide linked together via a single disulfide bond. The main sequence
in the heavy chain is due to the alanine-glycine repeats. Fibroin has higher
proportions of glycine and alanine acids. Fibroin has water-absorbing,

insulating, and lustrous nature. Thermotolerance is yet another noticeable property of fibroin [6, 11] (Figure 5.2).

FIGURE 5.1 Components of silk.

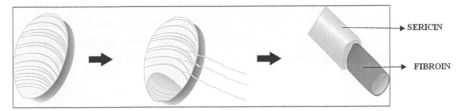

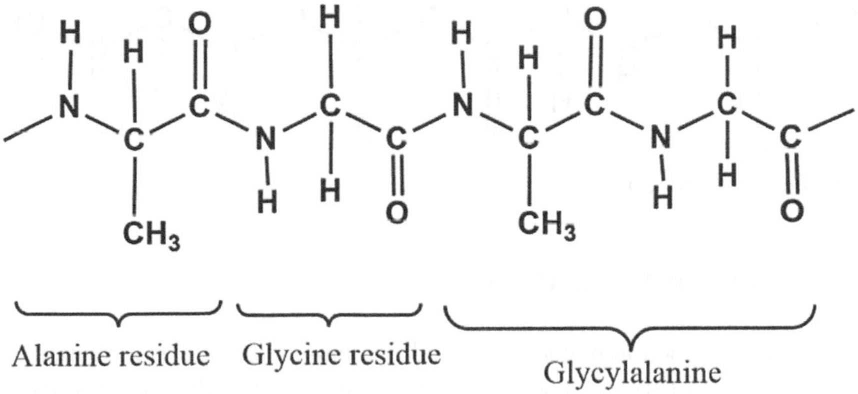

Alanine residue   Glycine residue   Glycylalanine

FIGURE 5.2 Structure of silk fibroin.

5.2.2 SILK SERICIN

Sericin is yet another constituent of silk which provides adhesion to the fibroin in cocoon. It was considered a waste material in the textile industry until recent past. After the properties of the sericin were closely observed, it found a lot of applications in biomedical field. It is very easily hydrolyzed. It becomes a user due to its specific properties like resistance to oxidation, antibacterial activity. They also possess UV resistance and absorb/ release moisture efficiently, inhibitory activity. It is carefully separated from fibroin during the silk manufacturing process, to make silk appearance lustrous. The sericin was typically considered waste until these properties of it were known [5]. The chemical structure and molecular weight of sericin depend on the method of separation of sericin and fibroin and process of recovering sericin from degumming. Sericin is composed of more with serine and aspartic acids

[6]. Sericin has been to have properties to become hydrogel with sustained-release property, pH-dependent degradation dynamics, and good elasticity. It is also reported in literature that it has no immunogenic responses by various reporters [11]. Zhang et al. [40] reported that sericin protein could be cross-linked, copolymerized, and blended with other macromolecular materials, to make materials with enhanced properties. They also reported its usage, as an improved coating material for natural and artificial fibers, fabrics, and articles (Figure 5.3).

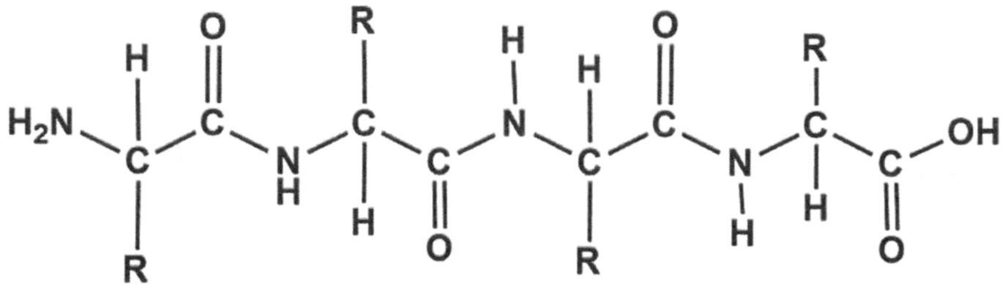

FIGURE 5.3   Structure of sericin.

5.3   SILK IN BIOMEDICAL FIELD

Silk has undergone many chemical modifications to emerge as a biomedical tool in various forms, such as silk solution, films, scaffolds, particles, nano-fibers, hydrogels, etc. Owing to its biocompatibility, slow biodegradability, controllable structure and morphology, self-assembly, and excellent mechanical properties silk has contributed widely towards the biomedical field. Silk is less inflammatory than other conventional biodegradable polymers. Among this biocompatibility and biodegradation are the major ones Silk is less inflammatory than other conventional biodegradable polymers and also found to have less or no immunogenic response unless fibroin and sericin are compared [11].

5.3.1   DRUG DELIVERY

Drug delivery is the method of administering a medicated compound to achieve a therapeutic effect in the body. The controlled release of the drugs is at the target size is a major concern in treatment. The literature reports numerous applications of silk proteins to have in delivering drugs at a controlled rate. The prominent processes like Gelation, microparticle, freeze-drying, electrospinning, electrospraying, etc. are the primary techniques explored in this area [11].

DadrasChomachayi, Masoud et al. reported the sustained drug release by SF and gelatin (GT) nanofibrous substrate thyme essential oil (TEO) and doxycycline monohydrate (DCMH), both as anti-bacterial agents, were loaded and studies were conducted. Both of them showed excellent cell-compatibility. The addition of GT has helped improving hydrophilicity, surface wettability, and a mass loss percentage of silk [8]. Suktham, Kunat et al. successfully attempted to load resveratrol to silk protein (SP) nanoparticles via a solventless precipitation method. Cell viability assay proved that Resveratrol-loaded SP nanoparticles could effectively inhibit the growth of colorectal adenocarcinoma (Caco-2) cells. Although the advantages found were that, it was non-cytotoxic to skin fibroblasts [9]. Gao Ya et al. has studied a transdermal drug delivery with SF, where silk acted as a scaffold to adsorb prepolymer solution and initiate microneedles formation of PEGDA/sucrose. It was successful in transdermal delivery and controllable release of Rhodamine B, indocyanine green, and doxorubicin by regulating the sucrose content. SF aided the hasty fabrication of polymeric microneedles for transdermal drug delivery applications [10]. Gupta et al. prepared SF-encapsulated with curcumin nanoparticles less using the devised capillary-microdot technique for drug delivery applications [12]. The drug encapsulation efficiency of various nanoparticles like pramipexole, curcumin, and propranolol hydrochloride on encapsulation with SF was studied by Gianak, Olga et al. The studies revealed among solvents propranolol showed the highest effect [13]. Zhuping Yin et al. synthesized swellable microneedles with SF using polydimethylsiloxane (PDMS). They were quickly penetrable to porcine skin with a depth of ~200 μm in vitro, and transform into semi-solid hydrogels and form 50–700 nm porous networks inside [14].

5.3.2 WOUND DRESSING

A wound is biologically defined as a fracture in healthy body integrity. Therefore, repairing the wound comprises the activation of intracellular and intercellular pathways for the restoration of this tissue integrity and homeostasis [16]. Literature reports SF-based dressings being used as carriers for, growth factors, delivering drugs, and bioactive agents to the wound area, along with appropriate aid for complete healing [15–20]. Antimicrobial properties imparting dressings have been traditionally used to minimalize the bacterial infection of burns and wounds. Meanwhile, Silk has been used in suturing for centuries.

Uttayarat, Pimpon et al. manufactured SF mats coated with silver nanoparticles (AgNPs) to act as a prototypic wound dressing and also evaluated for antimicrobial properties. The colloidal AgNPs were prepared by irradiating silver nitrate by gamma rays which were confirmed by transmission electron microscopy. The growth of *Staphylococcus aureus* and *Pseudomonas aeruginosa* was effectively inhibited on the application of this [15]. Karahaliloglu, Zeynep et al. reported the usage of chitosan/silk sericin (CHT/SS) scaffolds collective with lauric acid (LA) and zinc oxide (ZnO) nanoparticles for the successful wound dressing applications [16]. Mehrabani, MojtabaGhanbari et al. synthesized an interconnected microporous structured wound dressing material from chitin/SF/TiO$_2$ nanocomposite which has high antibacterial, blood clotting, and mechanical strength properties using the freeze-drying method. The studies proved that swelling and water uptake of the dressing was 93%, which highly contributes to wound dressing applications. Hemostatic potential studies showed that blood clotting ability of the nanocomposites was high on comparing with pure components and commercially available products.

Moreover, cell viability, attachment, and proliferation were confirmed by MTT assay and DAPI staining [17]. A lot of other natural dopant incorporated silk has also been the focus of researches these days [18–20]. Selvaraj et al. included fenugreek, a natural antioxidant to SF to form nanofiber by a co-electrospinning method. The biocompatibility and antioxidant activity were evaluated were found to be increased and were confirmed through MTT assay and DPPH scavenging assay, respectively. The experiment has proven that on incorporation of fenugreek thermal and mechanical properties of SF nanofibers has increased [20]. Sericin has also proven its wound dressing capability. Huawei He et al. reported the sericin/polyvinyl alcohol (PVA) blend film incorporated with AgNPs to have functional antibacterial activities against *E. coli* and *S. aureus* [21].

Lamboni, Lallepak et al. have reported the functionalization of bacterial cellulose using silk sericin for proper wound healing [23]. Hence, silk has become a successful biomimetic candidate for wound dressing and potential replacements for treating skin injuries [22].

5.3.3 TISSUE ENGINEERING (TE)

Tissue engineering (TE) is an interdisciplinary field in which engineering of tissues and organs by integrating science and technology of cells, materials, and biochemical factors is carried out. The most challenging part is the replication of the natural extracellular matrix (ECM). Hence, scaffold engineering is the prime concern in this field. Amongst the variety of materials tested, SF has proven as a promising material for this. SF has been reported as films, porous matrices, hydrogels, nonwoven mats, etc., for various TE applications in bone, tendon, ligament, cartilage, skin, liver, trachea, nerve, cornea, eardrum, dental, bladder [23–27]. Biman B. Mandal et al. have prepared novel 3-D sericin/GT scaffolds and 2-D films with uniform pore distribution, improved compressive strength, high swell ability, and high overall porosity all of which support TE and biomedical applications. The enhanced cell attachment, viability, and low immunogenicity were also reported which proved its potential as a future biopolymeric graft material [27]. Chao, Pen-Hsiu Grace et al. have reported the successful attempts of silk hydrogels in cartilage TE [25]. Lawrence, Brian et al. said the application of silk in cornea TE [28]. Kim, Hyeon Joo et al. reported the biomimetic growth of calcium phosphate on porous SF polymeric scaffolds to generate organic/inorganic composites as scaffolds for bone TE. Hence prepared mineralized protein-composite scaffolds were subsequently seeded with human bone marrow stem cells and found to be successful [29]. Nandana Bhardwaj et al. fabricated three-dimensional (3D) blended scaffolds of silk and keratin by freeze-drying method and characterized their physicochemical, mechanical, and degradable properties extensively. SF and human hair keratin blended scaffolds were demonstrated as the promising dermal substitute for skin TE [30]. Manchineel, Shivaprasad et al. manufactured the SF/melanin composite films and electrospun fiber mats for skeletal muscle TE. The incorporation of melanin modulated the thermal stability and electrical conductivity of scaffolds. It has also checked the alignment of fibers in electrospun mats and provided excellent antioxidant properties to the scaffolds [31]. Franck et al. observed the ability of gel spun silk scaffolds in bladder TE and also its cell responses [41]. Despite all the signs of progress made, there are still fields unexplored in the TE application for silk.

5.3.4   MEDICAL IMPLANTS

The purpose of implantable devices is to imitate an organ and to replace an injured part to retain the normal functioning of the body. The most extensively used medical implants are heart, ears, eyes, knees, breasts, hips, bones, and cardiovascular system. Conventional materials like metals, ceramics, and synthetic polymers may lead to immunological rejection by the body and hence a replacement with biopolymer is always welcomed. Silk has been reported as biomedical implants in literature these days [32–37]. Mehrjou, Babak et al. fabricated homogeneous nanocones which were undergone oxygen plasma etching on the surface of silk films. This has increased surface energy [34]. SF is also found to be successful in dental implants [35]. Elia, Roberto et al. reported formulation of Risperidone by gelation of SF as injectable implants [36]. SF tubular scaffolds are found to be ideal in vascular graft implants. Various methods like dip coating, electrospinning has been reported in literature. Blends with polyethylene oxide are also found to be effective [37]. Lorenz Meinel et al. assessed the appropriateness of silk biomaterials as implants for bone healing. It was found that novel porous SF scaffolds permitted adequate temporal control of hydroxyapatite deposition and resulted in the formation of a trabecular-like bone matrix in the studies [38]. Altman et al. have developed a wire rope matrix for the expansion of autologous tissue-engineered anterior cruciate ligaments (ACL) using a patient's adult stem cells with silk [39]. Films made out of sericin and fibroin has excellent oxygen permeability it mimics the human cornea and its functional properties hence it is explorable in the future to make corneal implants out of silk [40].

5.4   ELECTROSPUN SILK IN BIOMEDICAL APPLICATIONS

Electrospinning is an evolving technique in synthesizing scaffolds and wound dressing materials using silk in biomedical field. Affordability, very high surface-to-volume ratio, variable porosity, and versatility in materials selection are the factors that make electrospinning a favorable one in biomedical applications. The ECM of human body is composed fibrous proteins that can be mimicked by the nanofibrous mat produced by electrospinning.

Li et al. reported the electrospinning of bioactive SF/hydroxyapatite (nHAp) to improving bone formation; he was successful in developing human bone marrow-derived mesenchymal stem cells on the scaffolds. Schneider et al. spun epidermal GF (EGF)-loaded silk nanofibers which could aid the healing process, thereby decreasing the time of wound closure [43].

Our research group has been involved in understanding the role of silk in elctrospun fibers of biodegradable polymers and their vast applications. We have carried out emulsion electrospinning of SF and polycaprolactone (PCL) for TE applications. The findings revealed various combinations of emulsion electrospinning rather than the conventional electrospinning of polymer blends. PCL has been dissolved in chloroform and SF in formic acid. The concentration of PCL was continuously taken at 10% (w/v), while that of SF concentration was varied. The physicochemical, mechanical, and interfacial properties of the nanofibrous electrospun mats were studied. Different combinations of the emulsion were prepared (Figure 5.4). The structure of SF and scaffolds were confirmed using FTIR (Figure 5.5). The resultant nanofibers exhibited a distinct core-shell structure without using co-axial electrospinning (Figure 5.6) [44]. These nanofibers could be a prospective vehicle for drug-delivery, growth factors, or biomolecule encapsulation in upcoming TE applications (Table 5.1).

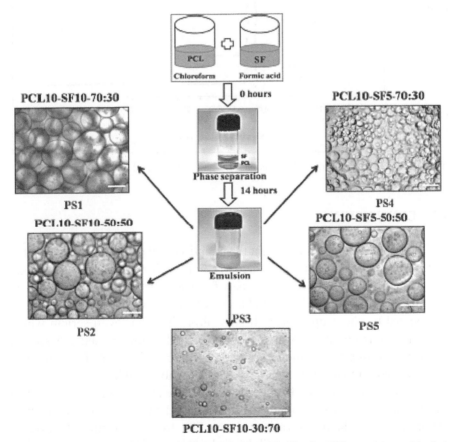

FIGURE 5.4   Different combinations of PCL and silk fibroin (SF) emulsion with their nomenclature, scale bar of optical images: 100 µm.

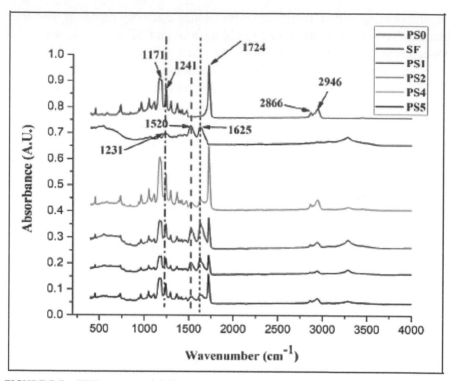

FIGURE 5.5   FTIR spectra of different nanofibrous scaffolds and SF powder.

5.5 CONCLUSION

Silk has been esteemed and caught the attention of the human race, by virtue of its lustrous appeal since long ago. Therefore, it rules the textile market across the globe even today. But apart from its aesthetic value researchers have attributed some medicinal values to silk in the recent past and its scope is expected to shoot up shortly. It was surprising that, even sericin which was once considered the waste material in silk industry had got special consideration due to its unique properties. Silk proteins were found to have excellent biocompatibility and slow biodegradability which provides it an unavoidable role in biomedical applications. The structure composition and current researches like wound healing, TE and the emerging biomedical applications of silk in its various forms like films, nanofiber, hydrogels, and particles were discussed. Silk by undergoing major and minor chemical modification or by combining with different synthetic or natural polymers can work wonders in the biomedical field. There is still a vast world of silk unexplored; the researchers have to research these hidden capable avenues and their applications.

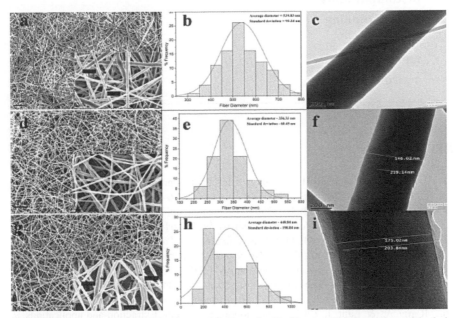

FIGURE 5.6 Different combinations of PCL and silk fibroin (SF) emulsion with their nomenclature, scale bar of optical images: 100 µM.
Scanning electron microscope (SEM) images of PS0 to PS5 nanofibers (a,d,g,j,m,p) (scale bar:10 µm) with magnified images of same (inset) (scale bar: 1 µm). Frequency distribution of nanofiber diameters of PS0 to PS5 b,e,h,k,n,q. Transmission electron microscope (TEM) images of PS0 to PS5 nanofibers (c,f,i,l,o,r).

TABLE 5.1   Properties of the Different Emulsion Combinations

Emulsion Combination	Nanofiber Name	Spinnability	Nanofiber Diameter
PCL 10%-SF 0%	PS0	Y	539.8 ± 93.2 nm
PCL 10%-SF 10%-70:30	PS1	Y	336.3 ± 60.4 nm
PCL 10%-SF 10%-50:50	PS2	¥*	448.8 ± 198.0 nm
PCL 10%-SF 10%-30:70	PS3	N	1021.6 ± 351.9 nm
PCL 10%-SF 5%-70:30	PS4	Y	212.6 ± 43.4 nm
PCL 10%-SF 5%-50:50	PS5	Y	370.1 ± 111.43 nm

Y – Yes; N – Not spinnable; ¥* – Spannable with secondary jets of nanofibers, random fiber.

KEYWORDS

- anterior cruciate ligaments
- biocompatibility
- electrospinning
- fibroin
- nanofibers
- sericin

REFERENCES

1. Leng-Duei, K., et al., (2015). Structures, mechanical properties, and applications of silk fibroin materials. *Progress in Polymer Science, 46*, 86–110.
2. Murphy, A. R., & David, L. K., (2009). Biomedical applications of chemically-modified silk fibroin. *Journal of Materials Chemistry,19*(36), 6443–6450.
3. Bloch, A., & Arthur, S. M., (1969). *Silk Suture*. U.S. Patent No. 3,424,164.
4. Pérez-Rigueiro, J., et al., (2000). Mechanical properties of single-brin silkworm silk. *Journal of Applied Polymer Science,75*(10), 1270–1277.
5. Mondal, M., (2007). The silk proteins, sericin, and fibroin in silkworm, *Bombyx mori* Linn.: A review. *Caspian Journal of Environmental Sciences,5*(2), 63–76.
6. Rangi, A., & Lalit, J., (2015). The biopolymer sericin: Extraction and applications. *J. Text Sci. Eng., 5*(1), 1–5.
7. Murphy, A. R., & David, L. K., (2009). Biomedical applications of chemically-modified silk fibroin. *Journal of Materials Chemistry,19*(36), 6443–6450.
8. Dadras, C. M., et al., (2018). Electrospun nanofibers comprising of silk fibroin/gelatin for drug delivery applications: Thyme essential oil and doxycycline monohydrate release study. *Journal of Biomedical Materials Research Part A,106*(4), 1092–1103.
9. Suktham, K., et al., (2018). Efficiency of resveratrol-loaded sericin nanoparticles: Promising bionanocarriers for drug delivery. *International Journal of Pharmaceutics,537*(1/2), 48–56.
10. Gao, Y., et al., (2019). Highly porous silk fibroin scaffold packed in PEGDA/sucrose micro needles for controllable transdermal drug delivery. *Biomacromolecules,20*(3), 1334–1345.
11. Dutta, D., Chowdhury, M. H., & Avijit, B., (2018). Silk proteins in drug delivery: An overview. *RPHS,4*(4), 514–518.
12. Gupta, V., et al., (2009). Fabrication and characterization of silk fibroin-derived curcumin nanoparticles for cancer therapy. *International Journal of Nanomedicine,4*, 115.
13. Gianak, O., et al., (2018). Silk fibroin nanoparticles for drug delivery: Effect of bovine serum albumin and magnetic nanoparticles addition on drug encapsulation and release. *Separations,5*(2), 25.

14. Yin, Z., et al., (2018). Swellable silk fibroin micro needles for transdermal drug delivery. *International Journal of Biological Macromolecules, 106,* 48–56.
15. Uttayarat, P., et al., (2012). Antimicrobial electro spun silk fibroin mats with silver nanoparticles for wound dressing application. *Fibers and Polymers,13*(8), 999–1006.
16. Karahaliloglu, Z., Ebru, K., & Emir, B. D., (2017). Antibacterial chitosan/silk sericin 3D porous scaffolds as a wound dressing material. *Artificial Cells, Nanomedicine, and Biotechnology,45*(6), 1172–1185.
17. Mehrabani, M. G., et al., (2018). Chitin/silk fibroin/TiO$_2$ bio-nanocomposite as a biocompatible wound dressing bandage with strong antimicrobial activity. *International Journal of Biological Macromolecules,116,* 966–976.
18. Elakkiya, T., et al., (2014). Curcumin loaded electro spun *Bombyx mori* silk nano fibers for drug delivery. *Polymer International,63*(1), 100–105.
19. Kheradvar, S. A., et al., (2018). Starch nanoparticle as a vitamin E-TPGS carrier loaded in silk fibroin-poly(vinyl alcohol)-Aloe vera nanofibrous dressing. *Colloids and Surfaces B: Biointerfaces,166,* 9–16.
20. Selvaraj, S., & Nishter, N. F., (2017). Fenugreek incorporated silk fibroin nanofibers a potential antioxidant scaffold for enhanced wound healing. *ACS Applied Materials and Interfaces,9*(7), 5916–5926.
21. He, H., et al., (2017). Preparation and characterization of silk sericin/PVA blend film with silver nanoparticles for potential antimicrobial application. *International Journal of Biological Macromolecules,104,* 457–464.
22. Farokhi, M., et al., (2018). Overview of silk fibroin use in wound dressings. *Trends in Biotechnology,36*(9), 907–922.
23. Lamboni, L., et al., (2016). Silk sericin-functionalized bacterial cellulose as a potential wound-healing biomaterial. *Biomacromolecules,17*(9), 3076–3084.
24. Kasoju, N., & Utpal, B., (2012). Silk fibroin in tissue engineering. *Advanced Healthcare Materials,1*(4), 393–412.
25. Pen-Hsiu, G. C., et al., (2010). Silk hydrogel for cartilage tissue engineering. *Journal of Biomedical Materials Research Part B: Applied Biomaterials,95*(1), 84–90.
26. Gupta, P., et al., (2016). Biomimetic, osteoconductive non-mulberry silk fiber reinforced tricomposite scaffolds for bone tissue engineering. *ACS Applied Materials and Interfaces8*(45), 30797–30810.
27. Mandal, B. B., Anjana, S. P., & Kundu, S. C., (2009). Novel silk sericin/gelatin 3-D scaffolds and 2-D films: Fabrication and characterization for potential tissue engineering applications. *Acta Biomaterialia,5*(8), 3007–3020.
28. Lawrence, B. D., et al., (2009). Silk film biomaterials for cornea tissue engineering. *Biomaterials,30*(7), 1299–1308.
29. Kim, H. J., et al., (2008). Bone tissue engineering with premineralized silk scaffolds. *Bone42*(6), 1226–1234.
30. Bhardwaj, N., et al., (2014). Silk fibroin-keratin based 3D scaffolds as a dermal substitute for skin tissue engineering. *Integrative Biology,7*(1), 53–63.
31. Manchineella, S., et al., (2016). Pigmented silk nanofibrous composite for skeletal muscle tissue engineering. *Advanced Healthcare Materials,5*(10), 1222–1232.
32. Rebelo, R., Margarida, F., & Raul, F., (2017). Biopolymers in medical implants: A brief review. *Procedia Engineering,200,* 236–243.
33. Elia, R., et al., (2015). Electrodeposited silk coatings for bone implants. *Journal of Biomedical Materials Research Part B: Applied Biomaterials,103*(8), 1602–1609.

34. Mehrjou, B., et al., (2019). Antibacterial and cytocompatible nanoengineered silk-based materials for orthopedic implants and tissue engineering. *ACS Applied Materials and Interfaces,11*(35), 31605–31614.

35. Ebrahimi, A., et al., (2018). Preparation and characterization of silk fibroin hydrogel as injectable implants for sustained release of risperidone. *Drug Development and Industrial Pharmacy,44*(2), 199–205.

36. Elia, R., et al., (2015). Silk electro gel coatings for titanium dental implants. *Journal of Biomaterials Applications,29*(9), 1247–1255.

37. Kundu, S., (2014). *Silk Biomaterials for Tissue Engineering and Regenerative Medicine* (p. 53). Elsevier.

38. Meinel, L., et al., (2005). Silk implants for the healing of critical size bone defects. *Bone,37*(5), 688–698.

39. Altman, G. H., et al., (2003). Silk-based biomaterials. *Biomaterials,24*(3), 401–416.

40. Yu-Qing, Z., (2002). Applications of natural silk protein sericin in biomaterials. *Biotechnology Advances,20*(2), 91–100.

41. Franck, D., et al., (2013). Evaluation of silk biomaterials in combination with extracellular matrix coatings for bladder tissue engineering with primary and pluripotent cells. *PloS One,8*(2), e56237.

42. Holland, C., et al., (2019). The biomedical use of silk: Past, present, future. *Advanced Healthcare Materials,8*(1).

43. Eatemadi, A., et al., (2016). Nanofiber: Synthesis and biomedical applications. *Artificial Cells, Nanomedicine, and Biotechnology, 44*(1), 111–121.

44. Roy, T., et al., (2018). Core-shell nanofibrous scaffold based on polycaprolactone-silk fibroin emulsion electro spinning for tissue engineering applications. *Bioengineering,5*(3), 68.

CHAPTER 6

Influence of Nanotechnology on Prevention of Biofouling in Aquaculture Cage Nets: Green Conducting Polymers and Their Significance in Case Studies

P. MUHAMED ASHRAF

ICAR Central Institute of Fisheries Technology, Cochin – 682029, Kerala, India, E-mail: ashrafp2008@gmail.com

ABSTRACT

Biofouling in aquaculture cages has been an important issue for the farmers since its management incurs a huge sum of money and labor. Antifouling strategies failed due to the diversity of organisms are different from region to region and increased concern about pollution in the water bodies. The chapter describes the strategies to combat biofouling using nanomaterials coated over a polyaniline coated polyethylene and its efficiency towards antifouling. Green synthesis of nanomaterial incorporated hydrogel and mixed charged polymeric hydrogel over polyethylene aquaculture cage nets against biofouling also described. Brief description of environmental impact assessment of nanomaterials treated nets in the marine environments included at the end.

6.1 INTRODUCTION: FISHERIES SCENARIO INDIA AND ABROAD

Growing populations in the world and demand for food security is increasing year by year. A part of the food security is met through marine and aquaculture produced fishery resources. India has 8041 km of coastline, 3 million hectares of reservoirs, and 1.2 million hectares of brackish water. India has shown incremental improvement in fish production and it provides employment to 14 million people. According to the National Fishery Development Board, report the fish production during 2017–2018 was 12.60 million tons of which 67% were from inland and culture production. This shows the contribution from the culture sector in the nutritional security is significant. Aquaculture production in India was second in the world and plays a major role in India, food security, labor, exports, and the economy [1]. Major aquaculture based fishes are carps, catfishes, prawns, tilapia, freshwater pearl culture, coldwater fisheries, and ornamental fish culture.

Dwindling fish catch, overexploitation, illegal, unreported, and unregulated (IUU) fishing, environmental issues, climate change (CC) and a labor shortage are major concerns regarding marine fishery has led to planners to concentrate on the aquaculture production [2–4]. Fish production through aquaculture is an absolute necessity to feed the growing population and fish is also considered as the cheapest protein source [5]. India produces fish 5% of the total world production and 7% of the total aquaculture production. In view of this, planners stressed the need to improve the aquaculture production by extending shallow waters to marine areas. Increased aquaculture production led to a series of related issues such as viral and bacterial diseases, deteriorated water quality, influence of climate change, unutilized nutrients, and organic wastes, increased input cost, low rate to the produce due to increased catch, escapement of exotic species in the surrounding waters, and finally the cost of maintenance of cages. Maintenance of the cages is a major problem and it cost about 25% of the project cost. The major issue in cages was biofouling and closure of lumens of meshes by fouler.

6.2   BIOFOULING IN UNDERWATER STRUCTURES

Submerged natural and artificial materials are used for the settlement of micro and macro-organisms are called biofouling. When a material is submerged in the marine or aquatic environment, initially the surface was absorbed by organic molecules like proteins, polysaccharides, and macro organic molecules. Followed by microorganisms like bacteria, diatoms, microalgae, and fungi are formed a complex nutrient-rich layer and the same layer was called biofilms [6–8]. Macro-organisms like mussels, barnacles, sponges, polychaetes, oysters, etc., are attached over this biofilm. These biofoulers form over the surfaces in parallel or overlapping mode and the attachment was strong and difficult to remove [9, 10].

6.2.1 AQUACULTURE

Aquaculture or aquafarming is the farming of fish, crustaceans, mollusks, algae, *aquatic organisms, and other aquatic organisms*. The farming implies, rearing of organisms under controlled conditions in fresh or saltwater to improve the growth, protection from predators, diseases, and enhance production. This was done open ponds, reservoirs, or farms. Cage culture implies the fishes reared in cages, exposed in the aquatic environment from fingerlings to table or marketable sizes. Cages are traditionally fabricated by different ways and it depends on the region and culture of the area. Commercial cage culture was practiced by fabricating circular or square cages comprised of polyethylene nets covered top and bottom. The cages are made upright by using floating frames, and sinkers. Intensive cage culture is coming up in India recent years and it was directly carried out marine environments (Figure 6.1).

FIGURE 6.1   Models of aquaculture cages in the coastal marine waters.

6.3   BIOFOULING AND ITS MANAGEMENT

Biofouling is the major nuisance in the aquaculture practices. Different types of marine organisms were attached over the cage nets and the lumens were clogged. This will prevent the water circulation, accumulation of wastes in the cages, nutrient accumulation, increase the weight of the cage, reduce the drag, and retard the growth of organisms in the cage. Biofouling also reduce the efficiency of the materials used for fabrication of cages by increasing weight, abrasion, and brittleness. Current practice is cleaning the cage meshes by mechanical or manual methods. The accumulation of biofoulers diversity and intensity will be varied with the seasons and geographic locations [11]. A common strategy to combat biofouling is very difficult since the species and nature of organism are different from places to places. A group of benthic organisms produce toxic metabolites which will affect the growth and life of organisms in the cages. Biofouling communities compete for resources with cultured fishes and may harm the fishes in the cages.

The cost of antifouling management in aquaculture cages are very expensive [12] and it was estimated about 25% of the total project cost. The cost involves periodic cleaning, replacement of nettings and pollution due to the cleaning process. This will make the farming become nonprofitable. The question how to mitigate the problem of biofouling in the aquaculture cages.

6.4   CURRENT APPROACHES TOWARDS ANTIFOULING

Biofouling is a serious problem in shipping industries and a large number of coating technologies were developed and employed to combat biofouling in ship hulls. Earlier days copper oxide-based antifouling paints, then tributyltin (TBT) and their analogous compounds were used to combat biofouling in ship hulls. TBT was banned due to environmental concerns. New generation antifouling coatings are mainly from low surface energy, self-polishing type paints. In fishing boats and other submerged materials such as pillars, dam shutters mainly, employed CuO-based antifouling paints. The basic principle behind the CuO-based paints are a specific amount of CuO was released to the environment at a fixed rate and this will be continued till concentration is minimum. In fishing boats, it will stay up to one year. Generally, CuO-based antifouling coating life is one year and they need to repaint. The world over aquaculture is getting prominence and the major issue raised by the farmers is biofouling and its management. Antifouling intervention in aquaculture system is very sensitive since it deals with organisms and food. It needs careful evaluation. So intervention using chemical methods needs lots of tests and safety approvals. A series of research attempts were made to combat biofouling in aquaculture systems.

The current scenario of management of aquaculture cages from fouling is mainly manual cleaning by brushing or mechanical scrubbers. Mechanical cleaning involves employing mechanical scrubbers, water jets, and scrappers. These are labor-intensive, pollute the cages and its vicinity by dispersing minute particles, disturbance to the organisms, and expensive. Other methods include air or sun drying by taking out the nets. This will not remove the fouler but kill the fouler and needs physical removal [13].

Popular antifouling strategy in aquaculture cages are by treating low-levelcopper-based chemical coating and a fixed quantity of biocide leaching the aquatic environment will deter the fouler. This treatment will last only for one season of culture. The cost of treating the net is very high. Costello et al. [14] developed a biodegradable copper oxide composite and these are not planned to use for aquaculture. Copper and tin-based molecules using against biofouling was banned by the EU and other countries [15]. Most of the copper-based antifouling coating over aquaculture cage nets are very heavily loaded with copper and some types of organisms will attach over these nets. Cleaning of these nets will create copper and sludge waste and difficult to dispose of. The other methods are:

• Biological control using grazers [16, 17];
• Silicon-based fouling-release coating [18];
• Acetic acid-based antifouling strategies over fish cage nets [19];
• Modifying the cage net design to avoid shellfish biofouling;
• Natural material based antifouling strategies [20, 21].

These are considered in laboratory-scale not much practical application under field conditions. Introduction of nanotechnology as potential tools for varying applications may be attempted for combating biofouling under submerged environments.

6.5 NANOTECHNOLOGICAL INTERVENTION

In the Japanese Society of Precision Engineering Conference, Japan, 1974, Prof. Taniguchi introduced the word nanotechnology, which means synthesis, fabrication, development of devices or materials having size 1 to 100 nm [22]. The principle of nanotechnology is those materials with known properties and functions will exhibit different behavior and functions at nano-sized state. By reducing, the size of the particles will increase enormously the surface area of the particle per unit weight. In other words per unit weight the number of nano-sized particles and surface area will be higher than the microstate. Generally, nanomaterial size will be varied between 1 to 100 nm. Nano-sized inorganic and carbon-based molecules are employed for solving the problems of drug delivery, sensors, catalysis, surface coatings, and anti-bacterial applications. Depends on applications different nanomaterials were employed. The advantage of nanotechnology is the use of very low amounts of materials, increased surface area, high activity, and high efficiency. Series of research undertaken to the application of nanomaterials against the use of antibacterial and biofouling applications.

Polyethylene is a non-polar molecule and its fibers used for the fabrication of aquaculture cage nettings. High-density polyethylene was extensively used for the fabrication of cage nettings (Figure 6.2). Due to its nonpolar nature difficult, to make any chemical-based interventions. As described earlier sections the antifouling strategies over polyethylene nets is difficult. ICAR CIFT is introduced as a new strategy to prepare biofouling resistant aquaculture cage nets by employing a different method [23]. Polyaniline is a conducting polymer and has good conductivity. Trials were conducted to coat the polyaniline over the net was failed. Then the polyaniline was synthesized in situ over polyethylene. The method involves the cage net was immersed in acidified aniline solution overnight and the ammonium persulfate was added slowly to the solution without disturbing the system. Dark greenish precipitates of polyaniline were formed and attached strongly over the surface. Due to immersion in acidic aniline solution, the surfaces were covered by adsorbed aniline. The slow polymerization process probably made the surface to attach a strong coating of polyaniline. The formation of polyaniline over the polyethylene examined through a scanning electron microscope (SEM) showed the nanorods of polyaniline were formed over the surface. The more nanorods of polyaniline over the surface higher will be the efficiency. Excess poly-aniline is removed by washing. Now the surface becomes conductive and ready to accept biocides over the surface. As stated earlier copper oxide is a known potential biocide against marine fouler. The concern raised against its use as biocide was due to the higher amount of biocide need to prevent biofouling and its principle is a fixed percentage of copper oxide continuously leached to the environment, this will deter the larvae of fouler from the surface. Application of nano copper oxide will reduce the amount used for prevention is from gram to milligrams or lesser. The minute amount will provide a higher biocidal effect. Hence, nano biocide is a good option to combat biofouling. The polyaniline coated cage nets were immersed in nano copper oxide dispersed aqueous solution overnight. The active copper oxide nanoparticles were adsorbed over the conductive polyaniline uniformly over the surface. The nano-sized particles easily penetrated to the interior layers of the polyaniline-polyethylene composite. The nanoparticles were having increased surface area, smallest size made them to adsorb over the surface, increased activity, and exhibit the efficient toxic effects. Now the material is ready for the field use.

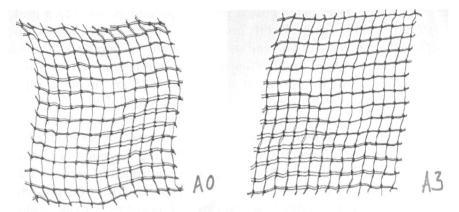

FIGURE 6.2   The aquaculture cage nets untreated and treated.

6.5.1   *SURFACE CHARACTERIZATION*

Surface of the cage nets were evaluated sophisticated techniques like SEM. The results clearly showed that the surface of the treated webbing was coated polyaniline and over that copper was adsorbed. Similarly, the atomic force microscope measurement exhibited uniform coating over the surface.

Fourier transform infrared (FTIR) spectroscopy measure the vibrational characteristics of the functional groups present in the molecule at it equilibrium position. The infrared light scanning from frequencies 400 to 4000 cm^{-1} will depict the characteristics of different functional group present in the molecules. The region 800 to 1800 cm^{-1} was known as fingerprint region and most of the functional groups will exhibit in this region. Few functional groups like OH, amines, CH_3, CH_2, and CH was generally identified from 2800 to 4000 cm^{-1}. The characteristic IR absorption frequencies of different functional groups will vary with the environment such as neighboring groups, or surrounding chemical bonds. Polyethylene has two types of CH groups viz., CH_3, CH_2 and their absorption was at 1460 ± 10 (symmetric stretching), 1360 ± 20 (Wagging), 2950 (symmetric str), 2850 (symmetric str), and 720 cm^{-1}. The polyethylene netting exhibited the IR absorption at frequencies 729, 1373, 1396, 1471, 1488, 2853, and 2924 cm^{-1}. The polyethylene was coated with polyaniline and IR spectra of the PE wagging doublet were shifted to 1362 and 1396 cm^{-1} indicating the polyaniline was coated over the polyethylene. The polyaniline will exhibit the specific quinonoid peak of NH_4^+/NH^+ at 1027/1141 cm^{-1}. The polyaniline formed over polyethylene exhibited peaks at 1047 /1181 cm^{-1} and nano copper oxide coated over the polyaniline showed a peak at 1070/1179 cm^{-1}. The peak shift from in 1027 to 1047 and 1070 clearly indicated the formation of polyaniline and adsorption of copper oxide over polyaniline respectively. Another functional group is benzenoid ring which was generally shows IR absorption at 1500 and 1585 cm^{-1} in polyaniline. The study showed the benzenoid ring absorption was shifted to 1508 and 1568 cm^{-1}. The characteristics of benzenoid ring were influenced due to the coating over polyethylene and adsorption of nano copper oxide. The broadening of the IR spectral peak implied that the formation of O-H-N or N---H-N bonding in the matrix. Specific IR absorption of copper oxide is 624 cm^{-1} which was shifted to 579 cm^{-1} when adsorbed over polyaniline. These results highlighted the formation of nano copper oxide coated polyaniline coating over polyethylene. The material needs to be tested its anti-biofouling efficiency in the laboratory or marine environments.

6.5.2   NANO CUO HYDROGEL-GREEN SYNTHESIS

Another approach towards the development of antifouling strategies is to use hydrogel. Hydrogels are considered as environmentally safe and its hydrophilicity will deter the adsorption of proteins. This will prevent the formation of biofilm, a prerequisite to attach microbes and fouler [24]. Hydrogels mainly used for development of antifouling strategies were from polyethylene glycol and their related compounds. These hydrogels alone under submerged condition loses its activity after some time. This leads to incorporate a biocide along with hydrogel. Ashraf [25] studied nano copper oxide incorporated PEG hydrogel was synthesized in situ over polyethylene aquaculture cage nets coated with polyaniline and the results showed that significant reduction in the accumulation of biofoulers after three months exposure in the Cochin estuary. The SEM and AFM micrograph showed the formation and uniformly coated nettings (Figure 6.3). The surface was uniformly covered with hydrogel and the treatment was 53% more efficient in controlling fouling attachment than untreated control. The nano copper oxide incorporated hydrogel showed the medium hydrophilicity than hydrogel alone coated nets. The order of hydrophilicity was hydrogel >CuO incorporated hydrogel > Untreated. The nano copper oxide incorporated hydrogel coating was semi hydrophilic, free from any defect, and more compact. The nano copper oxide in the hydrogel acted as a point source of biocide against microorganisms, which generally attacks the hydrogel.

FIGURE 6.3   SEM and AFM images of untreated and treated net samples.

6.5.3   MIXED CHARGED ZWITTERIONIC HYDROGEL

The hydrogels are not having much surface energy difference with surrounding waters and hence application slight thermodynamic force makes the fouler to stay permanently over the surface. Search for an alternative to lead to the new class of hydrogels called zwitterionic hydrogels. Zwitterionic polymers are a family of hydrogel which contain equal number of cations and anions. Generally synthesized by using sulfur, carboxy, or phosphor betaines. They are superhydrophobic in nature due to the presence of abundant ions and formation subsequent strong hydration layers. The protein and microorganisms were deterred by the betaine pendent structures. But these molecules are not very efficient in biofouling resistant. A new class of compounds called mixed charged copolymers which can resist biofouling as that of the zwitterionic hydrogel. Mohan and Ashraf [26] synthesized a nano-sized SiO_2-incorporated poly (Nisopropylacrylamide-co-2-(methacryloyloxy)ethyl]-Trimethylammonium/3-sulfopropyl methacrylate) mixed-charged zwitterionic polymeric hydrogel in situ over a polyethylene aquaculture cage netting material treated with polyaniline. The hydrogel was synthesized microwave method. The SiO_2 incorporated mixed charged polymeric hydrogel coated nets were exposed in marine environment exhibited biofouling resistance few months and it needs to modify to meet the industrial standards.

6.6   FIELD EVALUATION STUDIES

The pre-weighed treated aquaculture cage net and untreated cage nets were tied over PVC rack. The same is exposed in the sea by immersing it at 1 m from the surface through a floating platform. If it is exposing the estuarine environment, it must be immersed in such a way that 1 m down from the surface in non-tidal condition. The panels can be exposed for 1 to 9 months depending on the requirement of the researcher. After retrieval the samples were washed with running water to remove the sediments, loosely bound materials and other non-fouler. Allowed to dry in air and accumulation of biomass is noted (Figure 6.4). Analyze the diversity of fouling organisms by visual and microscopic evaluations. The diversity of fouler will be varied from season to season.

FIGURE 6.4   Field exposed samples untreated and treated.

6.7   LABORATORY EVALUATION

Ekbalad et al. [23] described in detail about the laboratory evaluation of biofoulers. The method describes the quantitative evaluation of the amount settlement of organisms and adhesion strength of the attached organisms. The testing materials were coated over an acid-washed clean glass slides and the same was used for different types of evaluations against fouler.

6.7.1   BIOFOULING ORGANISMS

One of the major foulers under marine environment is barnacles. Cyprid settlement assays were done to assess the biofouling inhibition [27]. Three days old cyprids of about 20 Nos were added in 1 ml of artificial/natural seawater (ASW) over the biocide coated glass slides (about 10 replicates). The cyprids were incubated at 28°C in dark for 48 h. After 24 and 48 h, the slides were taken out and enumerated the cyprid settlement and calculated the mean percentage settlement. Krusal-Wallis test and related statistical tests to be performed for evaluating statistical significance.

6.7.2   MICRO ORGANISM

Callow et al. [28] described the assays for the settlement of Ulva zoospore over the treated surface of glass plate in Quadriperm dish (Greiner Bio-one Ltd). 10 ml of suspension containing 1.5×10^6 mL^{-1} zoospores were added to the dish kept in dark for 45 min. Take out the slides and wash to remove unsettled swimming zoospores. The slides were fixed in 2.5% (v/v) glutaraldehyde, washed, and air-dried. The density of settled organism was measured through microscope as described by Callow et al. [29]. The adherence strength of zoospores can be estimated by exposing the slides to a wall shear stress of 52 Pa in a calibrated flow channel. The details will be available in Schulz et al. [30]. The data expressed in percentages.

Diatom assays can be carried out by using cells of *Navicula perminuta* [31]. It is resuspended in artificial seawater having chlorophyll-a concentration 0.30 µg ml^{-1} and the treated surfaces were placed in Quadriperm dishes. 10 ml of diatom suspension was added into it. The slides were exposed for 2 h and removed. The unattached diatoms were eliminated by washing and the slides were examined through a microscope as described above.

Antibacterial influence of the treated biocide was tested by using two marine bacteria species viz. Cobetia marina and *Marinobacterhydrocarbonoclasticus*. The former represents hydrophilic and the latter is hydrophobic in nature. C marina was used earlier as a model organism as it was considered bacteria of initial attachment in marine environment [32, 33]. The ASW conditioned (2 h) biocide coated slides were exposed in 8 ml bacterial suspension with standardized absorption of 0.2 AU at 595 nm for 1 h. the experiment was done polystyrene (PS) quadriperm dish. The slides were removed and washed to remove unattached bacteria. Then suspend the slide again quadrperm dish containing 8 ml sterile filtered seawater with fortified growth medium. Incubate the slides for 4 h at 30°C. Removed slides were stained using fluorochrom SYTO 13 and quantified using Tecan plate reader.

6.8   CONCLUSION

Biofouling in aquaculture cage nets an important problem and its management is very expensive. Biofouling organism's nature and diversity is different from marine, estuary, lakes, and region. Currently, not many efficient technologies are available. Application of nanomaterials based intervention can offer better antifouling management since it will not affect much the materials and organisms. The concentration of nanomaterials is very meager and the materials like copper will easily undergo complexation with water-soluble salts.

KEYWORDS

- **aquaculture**
- **artificial/natural seawater**
- **biofouling**
- **nanomaterials**
- **polyaniline**

REFERENCES

1.   Sam, S., Maheswaran, M., & Gunalan, B., (2015). Indian seafood industry strength weakness, opportunities, and threat in the global supply chain. *International Journal of Fisheries and Aquatic Studies, 3*(2), 199–205.
2.   Salagrama, V., (2004). *Policy Research: Implications of Liberalization of Fish Trade for Developing Countries: A Case Study for India.* Unpublished report. Kakinada: ICM.
3.   FAO, (2014). *The State of World Fisheries and Aquaculture-Opportunities and Challenges.*
4.   Shyam, S. S., & Narayanakumar, R., (2012). *World Trade Agreement and Indian Fisheries Paradigms: A Policy Outlook.*
5.   FAO, (2004). *The State of World Fisheries and Aquaculture* (p. 153). Rome FAO.
6.   Dobretsov, S., Teplitski, M., & Paul, V., (2009). Mini-review: Quorum sensing in the marine environment and its relationship to biofouling. *Biofouling, 25*(5), 413–427.
7.   Jain, A., & Bhosle, N. B., (2009). Biochemical composition of the marine conditioning film: Implications for bacterial adhesion. *Biofouling, 25*(1), 13–19.
8.   Avelelas, F., Martins, R., Oliveira, T., Maia, F., Malheiro, E., Soares, A. M., Loureiro, S., & Tedim, J., (2017). Efficacy and ecotoxicity of novel anti-fouling nanomaterials in target and non-target marine species. *Marine Biotechnology, 19*(2), 164–174.

9. Qian, P. Y., Lau, S. C., Dahms, H. U., Dobretsov, S., & Harder, T., (2007). Marine biofilms as mediators of colonization by marine macro organisms: Implications for antifouling and aquaculture. *Marine Biotechnology, 9*(4), 399–410.

10. Yebra, D. M., Kiil, S., & Dam-Johansen, K., (2004). Antifouling technology-past, present, and future steps towards efficient and environmentally friendly antifouling coatings. *Progress in Organic Coatings, 50*(2), 75–104.

11. Lane, A., & Willemsen, P., (2004). Collaborative effort looks into biofouling. *Fish Farming Int., 44,* 34–35.

12. Beaz, D., Beaz, V., Dürr, S., Icely, J., Lane, A., Thomason, D., Watson, P., & Willemsen, P. R., (2005). Sustainable solutions for mariculture biofouling in Europe. In: *ASLO Conference, Santiago Da Compostela.* Spain. Home page address: http://www. crabproject.com/index.php/57/publications (accessed on 3 August 2020).

13. Arakawa, K., (1980). *Prevention and Removal of Fouling on Cultured Oysters: A Handbook for Growers.* Maine Sea Grant Technical Report No. 56.

14. Costello, M. J., Grant, A., Davies, I. M., Cecchini, S., Papoutsoglou, S., Quigley, D., & Saroglia, M., (2001). The control of chemicals used in aquaculture in Europe. *J. Appl. Ichthyol., 17,* 173–180.

15. IMO, (2002). Biocides Directive EC 98/8/EC (European Commission. 1988).

16. Hidu, H., Conary, C., & Chapman, S. R., (1981). Suspended culture of oysters: Biological fouling control. *Aquaculture, 22,* 189–192.

17. Lodeiros, C., & García, N., (2004). The use of sea urchins to control fouling during suspended culture of bivalves. *Aquaculture, 231*(1–4), 293–298.

18. Baum, C., Meyer, W., Fleischer, L. G., & Siebers, D., (2002). *Biozidfreie Antifouling Beschichtung.* EU Patent EP1249476A2.

19. Carver, C. E., Chisholm, A., & Mallet, A. L., (2003). Strategies to mitigate the impact of *Ciona intestinalis* (L.) biofouling on shellfish production. *Journal of Shellfish Research, 22*(3), 621–631.

20. McCloy, S., & De, N. R., (2000). Novel technologies for the reduction of biofouling in shellfish aquaculture. In: Fisheries, N., (ed.), *Flat Oyster Workshop* (pp. 19–23). Sydney.

21. De, N. P. C., Steinberg, P. D., Charlton, T. S., & Christov, V., (2004). *Antifouling of Shellfish and Aquaculture Apparatus.* U.S. Patent 6,692,557.

22. Tret'yakov, Y. D., (2007). Challenges of nanotechnological development in Russia and abroad. *Herald of the Russian Academy of Sciences, 77*(1), 15–21.

23. Ashraf, P. M., Sasikala, K. G., Saly, N. T., & Leela, E., (2017). Biofouling resistant polyethylene cage aquaculture nettings: A new approach using polyaniline and nano copper oxide. *Arabian Journal of Chemistry.* http://dx.doi.org/10.1016/j. arabjc.2017.08.006 (accessed on 3 August 2020).

24. Ekblad, T., Bergström, G., Ederth, T., Conlan, S. L., Mutton, R., Clare, A. S., Wang, S., et al., (2008). Poly(ethylene glycol)-containing hydrogel surfaces for antifouling applications in marine and freshwater environments. *Biomacromolecules, 9*(10), 2775–2783.

25. Ashraf, P. M., (2019). Nano CuO incorporated polyethylene glycol hydrogel coating over surface modified polyethylene aquaculture cage nets to combat biofouling. *Fishery Technology, 56*(2019), 115–124.

26. Mohan, A., & Ashraf, P. M., (2019). Biofouling control using nano silicon dioxide reinforced mixed-charged zwitterionic hydrogel in aquaculture cage nets. *Langmuir, 35*(12), 4328–4335.

27. Hellio, C., Marechal, J. P., Veron, B., Bremer, G., Clare, A. S., & Le, G. Y., (2004). Seasonal variation of antifouling activities of marine algae from the Brittany coast (France). *Marine Biotechnology, 6*(1), 67–82.
28. Callow, M. E., Callow, J. A., Pickett-Heaps, J. D., & Wetherbee, R., (1997). Primary adhesion of enteromorpha (chlorophyta, ulvales) propagules: Quantitative settlement studies and video microscopy 1. *Journal of Phycology, 33*(6), 938–947.
29. Callow, M. E., Jennings, A. R., Brennan, A. B., Seegert, C. E., Gibson, A., Wilson, L., Feinberg, A., et al., (2002). Micro topographic cues for settlement of zoospores of the green fouling alga enteromorpha. *Biofouling, 18*(3), 229–236.
30. Schultz, M. P., Finlay, J. A., Callow, M. E., & Callow, J. A., (2000). A turbulent channel flow apparatus for the determination of the adhesion strength of micro fouling organisms. *Biofouling, 15*(4), 243–251.
31. Pettitt, M. E., Henry, S. L., Callow, M. E., Callow, J. A., & Clare, A. S., (2004). Activity of commercial enzymes on settlement and adhesion of cypris larvae of the barnacle *Balanus*amphitrite, spores of the green alga *Ulva*linza, and the diatom Navicula perminuta. *Biofouling, 20*(6), 299–311.
32. Akesso, L., Pettitt, M. E., Callow, J. A., Callow, M. E., Stallard, J., Teer, D., Liu, C., et al., (2009). The potential of nano-structured silicon oxide type coatings deposited by PACVD for control of aquatic biofouling. *Biofouling, 25*(1), 55–67.
33. Ista, L. K., Fan, H., Baca, O., & López, G. P., (1996). Attachment of bacteria to model solid surfaces: Oligo (ethylene glycol) surfaces inhibit bacterial attachment. *FEMS Microbiology Letters, 142*(1), 59–63.

CHAPTER 7

Biodegradable PHAs: Promising "Green" Bioplastics and Possible Ways to Increase Their Availability

TATIANA GR. VOLOVA,[1,2] EVGENIY G. KISELEV,[1,2] ALEKSEY V. DEMIDENKO,[1,2] SVETLANA V. PRUDNIKOVA,[1] EKATERINA I. SHISHATSKAYA,[1,2] and SABU THOMAS[1,3]

[1] *Siberian Federal University, 79 Svobodnyi Av., Krasnoyarsk – 660041, Russia, E-mail: volova45@mail.ru (T. G. Volova)*

[2] *Institute of Biophysics SB RAS, Federal Research Center "Krasnoyarsk Science Center SB RAS," 50/50 Akademgorodok, Krasnoyarsk – 660036, Russia*

[3] *International and Interuniversity Center for Nano Science and Nano Technology, Mahatma Gandhi University, Kottayam, Kerala, India*

ABSTRACT

Two approaches have been investigated in order to reduce the cost and increase the availability of the use of biodegradable polymers of microbiological origin (polyhydroxyalkanoates, PHA): (1) expanding the raw material base and attracting glycerol as a carbon substrate (large-tonnage waste from the production of biodiesel); and (2) filling the polymer with natural materials for industrial use. An effective technology for the synthesis of destructible polymers was developed using a new and more affordable substrate compared to sugars-glycerol. The *Cupriavidus eutrophus* B-10646 strain was studied as a product of polymers on glycerin, and the kinetic and production characteristics of the strain providing polymer yields on glycerol of various cheat, comparable with the process on sugars, were studied. Replacing sugars with glycerol provides a reduction in the cost of a carbon substrate in the production of PHA. To increase the availability of PHA and their technical applications, mixtures of PHA with natural materials were formed, the structure and properties, as well as the patterns of degradation in the soil, were studied. The possibility of long-term functioning of mixed forms in the soil with a gradual release of active substances (herbicides and fungicides) and the possibility of providing plants with protective equipment during the growing season are shown. Both approaches are effective for enhancing accessibility and expanding the scope of application of perspective green plastics-destructible PHA.

7.1   INTRODUCTION

The concept of sustainable development is the basic idea of the 21st century. Annual production of synthetic plastics has exceeded 320 million tons, and they largely accumulate in landfills, occupying fertile arable lands. Plastic wastes ruin municipal sewage and drainage systems. Up to 10% of the annual plastic production leaks into the Global Ocean [1]. New methods of economic management should introduce new, functional, and environmentally friendly recyclable materials in industrial production of the consumption of nonrenewable fossil resources, preserving them for future generations, enable more effective use of energy resources. Therefore, development of new environmentally friendly materials, which will completely degrade without releasing toxic products, should be the priorities for technologists of the 21st century.

Industrial ecology and green chemistry are tools for creating new eco-friendly materials including those manufactured from renewable sources [2]. Hybrid materials based on synthetic polyolefins and natural materials have been increasingly used in various applications. Polymers synthesized by living systems, the so-called biopolymers, have attracted considerable attention, too. Among various natural materials, a special place is occupied by polyhydroxyalkanoates (PHAs)-microbial polymers that can be used in different areas. PHAs are degraded to harmless products (CO_2 and H_2O); they are biocompatible, mechanically strong, thermoplastic, capable of blending with other materials, and processable by conventional methods from different phase states [3–7].

To a large extent (up to 45–48%), the cost of PHAs is determined by the cost of high purity carbon substrates [8, 9]; therefore, one of the most high-priority areas of research is the development of technologies using low-cost substrates.

However, the high cost hinders the use of PHAs in technical applications such as agriculture (films and pots for greenhouses, carriers for fertilizers and agrochemicals); production of packages, containers, and household goods; construction industry; etc., which require very large amounts of material. PHA cost can be reduced by upgrading production processes and by using cheaper substrates. Another approach is to use PHA blends with cheaper materials: this can both reduce the cost of the material and modify the properties of the polymers.

There are few published data on production and properties of PHA blends with solid-phase fillers. Some relatively recent studies investigated natural materials as fillers for modifying properties of PHAs. They described results of filling PHAs with clay and its derivatives [10], plant fibers [11], lignin, and holocellulose from a lignocellulosic biowaste [12], and wood chips and powder [13, 14]. As those studies describe different methods of production and investigation of PHA-based blends, it is difficult to compare them and draw any conclusion about the consistent effects of the fillers on the properties of these polymers. So, obtaining composite materials based on PHA is an unconditional resource for increasing their accessibility and expanding the scope of application.

Great potential for reducing the cost of PHA is the use of waste as raw materials. Potential raw materials for PHA synthesis are various substrates with different oxidation states of carbon, energy content, and cost, including individual compounds (carbon dioxide and hydrogen, sugars, alcohols, organic acids) and byproducts of alcohol and sugar industries, chemical processing of plant raw materials, and production of olive, soybean, and palm oils [15–17]. PHAs can be synthesized from both individual carbon compounds and various industrial wastes, making them economically feasible materials for different applications. Thus, the cost-effectiveness of PHAs can be enhanced by using new strains and carbon sources and by improving manufacturing processes, including large-scale ones. Scaling up laboratory biotechnologies to pilot production (PP) is a necessary step towards commercial production.

One of the promising substrates for large-scale production of PHAs is glycerol, the scale of production of which is currently increasing. This is due to the growing production of biodiesel as an alternative renewable energy source [18]. Biodiesel production increased dramatically from 500,000 gallons in 1999 to 450 million gallons in 2007, and it has been growing steadily over the past few years. In 2010, the global production of biodiesel reached 19 billion L [19], rising to more than 28 billion L in 2017 [20].

Glycerol is a by-product, amounting to about 10%, of biodiesel production by transesterification of animal and vegetable fats and oils (rapeseed, mustard, soybean, and palm oils) [15]. In industrial glycerol grades, the water content varies between 5.3 and 14.2%; methanol content - 0.001-1.7%; NaCl-traces-5.5%; K_2SO_4-0.8–6.6%. The analysis of publications generally indicates that glycerol has the potential to be used as a substrate for PHA production. However, the presence of impurities in glycerol indirectly affects the synthesis of PHA and their properties. It is obvious that involving new strains and upgrading the technological stages of the process will contribute to the improvement of PHA production.

In this work, our results are presented that are aimed at increasing the availability of PHA, which requires the use of glycerol as a substrate [6, 21–23], as well as the preparation of PHA composites with available natural materials [24].

7.2 MATERIAL AND METHODS

The strains used in this study were *Cupriavidus eutrophus* B-10646, registered in the Russian Collection of Industrial Microorganisms (RCIM). Schlegel's mineral medium was used as a basic solution for growing cells: $Na_2HPO_4 \cdot H_2O$ – 9.1; KH_2PO_4 – 1.5; $MgSO_4 \cdot H_2O$ – 0.2; $Fe_3C_6H_5O_7 \cdot 7H_2O$ – 0.025; $CO(NH_2)_2$ – 1.0 (g/L). Nitrogen was provided in the form of urea, and, thus, no pH adjustment was needed. The pH level of the culture medium was stabilized at 7.0 ± 0.1.

To cultivate bacteria in shake flask culture, an Innova 44 constant temperature incubator shaker (New Brunswick Scientific, U.S.) was used. Inoculum was prepared by resuspending the reference bacterial culture maintained on agar medium. The reference culture was grown in 1.0–2.0 L glass flasks half-filled with liquid saline medium, with the initial concentration of glycerol from 5 to 10 g/L. Growth kinetics of bacterial cells was studied in automated laboratory fermenters (Bioengineering AG, Switzerland), with 30-L and 150 L fermentation vessels and the working volume of the culture from 18 to 100 L, under strictly aseptic conditions.

Fermenters were equipped with systems for monitoring pH level, foam level, temperature, pressure, and dissolved oxygen. The fermenters were controlled by BioScadaLab software in automatic mode. To supply the feeding substrates, the fermenters were equipped with Bioengineering Peripex peristaltic pumps. The concentration of dissolved oxygen was maintained at

DO 30%. The air supply control was carried out in a cascade mode (DO-air flow-mixer revolutions). During cultivation in a 30-L fermenter, the amount of air supplied per cultivation process varied from 0 to 5.5 L/min, the speed varied from 500 to 1000 rpm. During cultivation in a 150-L fermenter, the amount of air supplied per cultivation process varied from 10 to 5.5 L/min, the speed varied from 300 to 750 rpm.

A two-stage process was used. In the first stage, cells were grown under nitrogen deficiency: the amount of nitrogen supplied in this stage was 60 mg/g cell biomass synthesized (i.e., 50% of the cell's physiological requirements – 120 mg/g); the cells were cultured in complete mineral medium and with glycerol flux regulated in accordance with the requirements of the cells. In the second stage, cells were cultured in a nitrogen-free medium; the other parameters were the same as in the first stage. The temperature of the culture medium was $30 \pm 0.5°C$ and pH was 7.0 ± 0.1.

During the cultivation, samples of culture medium were taken for analysis every 4–5 h (from fermenters) or every 8–10 h (from flasks); cell concentration in the culture medium was determined based on the weight of the cell samples dried at $105°C$ for 24 h (CDM); the total biomass (X_{total}) and catalytically active biomass (X_c) ($X_c = X_{total}$-PHA) g/L, were distinguished. Cell concentration in the culture medium was determined every hour by converting the optical density of culture broth at 440 nm to cell dry mass by using a standard curve prepared previously. The criteria for evaluating the process of PHA biosynthesis were as follows: concentration of cell biomass in culture, polymer content in cells, consumption of the main growth substrate, duration, and productivity of the process. The kinetic and production parameters of the culture were determined by conventional methods. The biomass concentration (X, g/L), the catalytic biomass concentration (X_c = X-PHA, g/L), the yield coefficient of the polymer (Y, g PHA/g substrate), the specific growth rate (m, h^{-1}), and the volumetric productivity (P, g/(L·h)) were calculated.

Intracellular polymer content at different time points was determined by analyzing samples of dry cell biomass. Intracellular PHA content and composition of extracted polymer samples were analyzed using a GC-MS (6890/5975C, Agilent Technologies, U.S.).

Three natural materials were used as fillers: peat, wood flour, and clay. High-moor peat "Agrobalt-N," state registration 0428-06-209-139-0-0-01, was produced by OOO "Akademiyatsvetovodstva," Russia). Wood floor was produced by grinding wood of birch (*Betula pendula* Roth) using an MD 250-85 woodworking machine ("StankoPremyer" Russia). Then it was dried at 60°C for 120 h until it reached constant weight, and 0.5 mm mesh was used to separate the particle size fraction. Clay was taken from the mine "Kuznetsovskoye," the Krasnoyarskii Krai, Russia. Its composition: loss on ignition – 6.92%; SiO_2 – 60.1%; Al_2O_3 – 19.17%; Fe_2O_3 – 6.72%; CaO – 2.02%; MgO – 2.12%; SO_3 – 0.65%; Na_2O – 0.88%; K_2O – 1.45%.

The polymer (poly-3-hydroxybutyrate) and fillers were pulverized by impact and shearing action in ultra-centrifugal mill ZM 200 (Retsch, Germany). To achieve high fineness of polymer grinding, the material and the mill housing with the grinding tools were preliminarily cooled at –80°C for about 30 min in an Innova U101 freezer (NEW BRUNSWICK SCIEN-TIFIC, U.S.). Grinding was performed using a sieve with 2-mm holes at a rotor speed of 18,000 rpm. Fillers were not precooled before grinding; a sieve with 2-mm holes was used, and the grinding was performed at 6000 rpm. Fractionation of the polymer and filler powders was carried out using vibratory sieve shaker AS 200 control (Retsch, Germany). Fractioning time was 10 min, amplitude 1.5 mm; a 100 g sample was used. Fractions of the polymer and filler powders of 200 μm and less were selected and used. Then, polymer powder was mixed with different filler powders in bench top planetary mixer SpeedMixer DAC 250 SP (Hauschild Eng., Germany); the blend time was 1 min, and the speed was 1000 rpm. The blends contained different amounts of the fillers: 10, 30, and 50%.

The homogenized blends were used to produce pellets and granules. Pellets were prepared by cold pressing at 36 bar using a laboratory semiauto-matic press (Minipress, Minsk, Belarus). Granules were prepared by mixing wet polymer paste (40%) in ethanol in planetary mixer SpeedMixer DAC 250 SP (Hauschild Eng., Germany); with the mixing time 1 min at 1000 rpm, and granulating in screw granulator Fimar EAC (Italy) with a 6 mm nozzle. The granules were dried in a fume hood at room temperature for 24 h.

Poly-3-hydroxybutyrate, peat, clay, and wood powders, powdered P(3HB)/filler blends, and granules and pellets were examined by using state-of-the-art physicochemical methods (SEM microscopy, FTIR spectroscopy,differential scanning calorimetry (DSC), and x-ray diffraction (XRD).

7.3   RESULTS AND DISCUSSION

7.3.1   *SYNTHESIS OF PHA ON GLYCEROL OF VARIOUS PURIFICATION*

Three brands of glycerin were used: Glycerol purified (Corporate Oleon, Sweden): glycerol – 99.3; chloride – 0.0001; salts (NH_4) – 0.005; Fe – 0.0005;

Ar − 0.00004; moisture − 0.09; fatty acid and ester − 0.25 (% mass); max heavy metal − 0.00005 µg/g (Glycerol I). Glycerol refinery B.V. (Dutch glycerol refinery, *Netherlands*): glycerol − 99.7; chloride <0.001; moisture − 0.09; fatty acid and ester − 1.0; sulfate <0.002; total organic impurities − 0.5–1.0; individual organic impurities − 0.1 (% mass); heavy metal <5 (µg/g) (Glycerol II). Crude glycerol (Prisma Comercial Exportadora de Oleoquimicos LTDA, Brazil): glycerol − 82.1; chloride − 4.35; mong − 0.13; methanol − 0.13; ash − 6.59; moisture − 9.88 (% mass); pH 5.8 (Glycerol III).

7.3.1.1   THE EFFECT OF GLYCEROL CONCENTRATION ON BACTERIAL GROWTH AND PHA SYNTHESIS

C. eutrophus B-10646 cells were cultivated in the media with concentrations of purified glycerol varied within a wide range to determine the limits of physiological action of this substrate for the strain and the kinetic constants [21]. The limits of physiological action of glycerol for this strain are very wide, varying between 0.5 and 60.0 g/L. There was a wide plateau (from 1 to 30 g/L), and the zones of limitation and inhibition of bacterial growth by glycerol were 0.1–3.0 and 30–60 g/L, respectively. The dependence of the specific growth rate (μ) on the substrate concentration (S) was described by the Andrews equation, which is a modified Monod equation. Using the graphical analysis method of Lineweaver-Burke ($1/\mu$:$1/S$) and Dickson's method ($1/\mu$:S), we calculated kinetic constants for this strain (saturation constant (Ks) and inhibition constant (Ki), μ_{max}). For the strain *C. eutrophus* B-10646, the limits of the physiological action of glycerol were found to be 1–30 g/L; K_s and K_i were 0.36 g/L (0.004 mol/L) and 62.0 g/L (0.673 mol/L), respectively; μ_{max} = 0.085 h⁻¹.

7.3.1.2   A STUDY OF THE GROWTH AND SYNTHESIS OF PHA BY C. EUTROPHUS B-10646 ON PURIFIED AND CRUDE GLYCEROL

The content of the main component (glycerol) in purified glycerol is more than 95–99%. In unpurified (crude) glycerol, depending on the raw material and the technology used, the content of glycerol is 80–85%; the other part comprises impurities, including free fatty acids (FFA) and methyl esters of FFA, alcohols, as well as water and salts, which, as a rule, inhibit the microorganisms responsible for PHA production.

The productivity of culture of *C. eutrophus*B-10646on different glycerol is shown in Table 7.1. Two grades of purified glycerol, with a 99.3 (Glycerol I) and 99.7% (Glycerol II) content of the basic material, and unpurified, crude glycerol (82.1%) were used in this study. Cultivation of *C. eutrophus* B-10646 cells was carried out in a 30-L fermenter with a starting cell concentration in the inoculate of 1.0–1.5 g/L, 25 g/L and 18 L working volume of the culture. The process was performed in a two-stage culture (30 hours) on saline medium: the growth of cells was limited by nitrogen deficiency in the first stage for 30–32 h; in the second stage (24–30 h), cells were grown in the nitrogen-free medium. The concentration of glycerol in the culture was maintained at 5–10 g/L, and it was fed into the culture with a peristaltic pump dispenser.

TABLE 7.1 Production Indicators of PHA Biosynthesis in a Culture of *C. eutrophus* B-10646 on Glycerol of Various Purification

Type of Glycerol	X, g/L	PHA, %	μ^*, ч$^{-1}$	Y_x	Y_p	Productivity (P), g/L*h	
						P_x	P_p
Glycerol-I	69.3 ± 3.5	72.1 ± 3.6	0.101	0.38	0.29	1.10	0.81
Glycerol-II	69.8 ± 3.5	73.3 ± 3.6	0.097	0.39	0.29	1.15	0.90
Glycerol-III	69.3 ± 2.9	73.3 ± 3.2	0.100	0.36	0.26	1.15	0.89

Source: Volova et al. [21].

Comparable cell concentrations and polymer contents were achieved in the experiments with *C. eutrophus* B-10646 cultivated on purified glycerol (Glycerol I and Glycerol II) (Table 7.1). When glycerol purified and glycerol refinery B.V. was used, the maximum yield of total biomass was 70 g/L, and the polymer content in cells was 72–75%. The analysis of the results revealed some differences in the kinetic parameters during the cultivation of bacterial cells. The specific growth rate of cells cultivated on glycerol purified was the highest (estimated from total and active biomass) in the initial period of the first stage of cultivation on complete nutrient medium with a limited supply of nitrogen (50% of the physiological requirement of bacteria), reaching 0.15 h^{-1} and 0.14 h^{-1}, respectively. This period corresponded to the most active consumption of glycerol by the culture, at an average specific rate of 4.0 ± 0.2 g/(g·h). The rate of polymer synthesis at this stage was 0.18 h^{-1}, and tended to decrease. After 30–32 hours, the total cell biomass concentration was 42.1 ± 1.7 g/L, and the polymer content in the cells reached $47.8 \pm 2.3\%$. During the second stage, the supply of nitrogen to the culture was stopped; the controlled supply of glycerol and mineral elements precluded the deficiency of these substrates in the culture. In the second stage, the specific growth rates of bacterial cells and polymer synthesis decreased gradually, as the consumption of glycerol dropped. At the end of the fermentation period (60 h), the total cell concentration was 69.3 ± 3.5 g/L, and the polymer content in the cells was $72.4 \pm 3.6\%$. The consumption of glycerol for the whole period was 3.5 ± 0.2 kg.

On the second type of purified glycerol (Glycerol II), the maximum values of the specific growth rate of bacteria (in terms of total and active biomass) in the first stage of the process were 0.15 h^{-1} and 0.13 h^{-1}, and they were comparable to the results for glycerol purified during this period. The consumption of glycerol in this period was between 3.8 and 4.0 g/(g·h)while the polymer synthesis rate was 0.18 ± 0.02 h^{-1} and tended to decrease. After 30 hours, the total cell biomass concentration was 45.6 ± 2.2 g/L, and the polymer content in the cells reached $56.1 \pm 2.7\%$. At the end of the fermentation period, the total cell biomass concentration was 69.4 ± 3.5 g/L, and the polymer content in the cells was $73.3 \pm 3.6\%$. The consumption of glycerol for the whole period was 3.4 ± 0.2 kg (Table 7.1).

On glycerol III, the maximum values of bacteria specific growth rate (in terms of total and active biomass) in the first stage of the process were 0.14 h^{-1} and 0.13 h^{-1}, respectively, when glycerol was consumed by the culture at a rate of 4.2 ± 0.2 g/g h. The average rate of polymer synthesis in this stage was 0.17 ± 0.02 h^{-1}. After 30 hours, the total cell biomass concentration was 46.2 ± 1.9 g/L, and the polymer content in the cells reached $52.2 \pm 2.1\%$. At the end of the fermentation period, the total cell biomass concentration was 69.3 ± 2.9 g/L, and the polymer content was $78.1 \pm 3.2\%$ (Table 7.1). The consumption of raw glycerol for the whole period amounted to 3.8 ± 0.2 kg, which corresponded to the yield coefficient for the polymer $- 0.26 \pm 0.01$, taking into account that the concentration of raw glycerol was 82.07%.

The consumption of raw glycerin for the entire period was 3.8 ± 0.2 kg, which corresponds to the economic coefficient for the polymer Yp $- 0.26 \pm 0.02$. Considering that its concentration in crude (raw) glycerin is 82.07%, the economic coefficient in terms of absolute glycerin is comparable to the values obtained with purified glycerin (Yp 0.29 ± 0.02 g/g).

Thus, it was shown that the production characteristics of *C. eutrophus* B-10646 when using glycerol of various purifications are comparable; the use of crude glycerol did not inhibit bacterial growth and PHA synthesis.

7.3.1.3   PILOT PRODUCTION OF PHA ON GLYCEROL

The process of PHA synthesis by *C. eutrophus* B-10646 on Glycerol II was scaled-up and studied under pilot production (PP) conditions [21, 22]. Pilot production consisted of units for media and inoculum preparation; a unit for fermentation; a unit for polymer extraction and purification. The PP fermentation unit included a steam generator (Biotron, South Korea) for sterilizing fermenters and service lines, a compressor (Remeza, Belarus) for air supply, a 30-L seed culture fermenter, a 150-L production fermenter, an ultrafiltration (UF) unit (Vladisart, Russia) to concentrate the culture, and a unit for cool dehumidification of the condensed cell suspension (LP10R ILSHIN C, South Korea).

The *C. eutrophus* B-10646 seed culture was derived from a reference stock culture maintained on agar medium, by growing cells in 2.0-L flasks on a complete nutrient medium in an incubator shaker. The resulting culture was concentrated by centrifugation in compliance with the sterility rules. Inoculum (4 L) with a cell concentration of 14–17 g/L was inoculated into a 30-L fermenter containing phosphate buffer. The process of building up the seed material was carried out on complete nutrient medium. For this purpose, dosing pumps were used to feed continuously separate flows of the solutions of glycerol, urea, magnesium sulfate, and ferric citrate with trace elements into the fermenter. The residual concentration of glycerol in the culture was maintained at 5–20 g/L; urea – 0.1–0.2 g/L; magnesium sulfate – 0.05–0.1 g/L.

The inoculum produced in a seed fermenter, was pumped into a 150-L production fermenter containing a sterile buffer solution of potassium and sodium phosphates in a sterile seeding line; thus, the initial concentration of the culture was 7.0 g/L. The process was carried out in two stages with continuous supply of sterile air and feeding solutions. At the end of fermentation (45 h), the cell biomass concentration was 110 ± 5.5 g/L, and the polymer content $78 \pm 3.1\%$. Thus, the average productivities of the polymer and cell biomass were 1.83 and 2.29 g/(L·h), respectively.

The results of the study of the chemical composition and physicochemical properties of PHA samples synthesized on glycerol of different purification degrees are presented in Table 7.2.

TABLE 7.2 Physico-Chemical Properties of Samples P(3HB) Synthesized on Glycerol of Various Purities

Substrate	M_w, кDa	D	C_x, %	T_g, °C	T_c, °C	T_{melt}, °C	T_{degr}, °C
Glucose	920	2.52	76	-	92	178	295
Glycerol I	355	3.42	50	2.9	96	174	296
Glycerol-II	416	3.63	55	-	103	176	296
Glycerol-III	304	3.49	52	2.7	99	172	295

"–" means no peak.

Source: Volova et al. [21, 22].

The properties of PHA samples synthesized by *C. eutrophus* B-10646 on three types of glycerol did not differ dramatically. The polymers synthesized by the glycerol-adapted productive culture *C. eutrophus* B-10646 on two types of purified glycerol had similar values of Mn-104 and 115 kDa, Mw-355 and 416 kDa, and polydispersity – 3.42 and 3.63, respectively. The Mn and Mw values of the PHAs produced on crude glycerol were somewhat lower, 87 and 304 kDa. These values were generally lower than those obtained earlier on other substrates. The ¹H-NMR spectra of three PHA samples synthesized on three glycerol grades were similar and showed the expected resonances for P(3HB) as demonstrated by the methyl group at 1.25 ppm, the methylene group between 2.45 and 2.65 ppm, and the methine group at 5.25 ppm. The ¹H-NMR-obtained spectra of PHA samples synthesized on glycerol were similar to those obtained by other researchers. The P(3HB) samples synthesized on glycerol had a reduced C_x (50–55%), i.e., the amorphous and crystalline regions had become nearly equal to each other. No deviations in the temperature properties were found in the samples of P(3HB). T_m and T_d values were within the previously identified value limits, 172–176, and 295–296°C, respectively.

It should be noted that the specific consumption of various substrates (sugars and glycerol) is comparable for the synthesis of PHA by the strain *C. eutrophus* B-10646. Table 7.3 presents a comparison of unit costs and the cost of various carbon substrates for the production of PHA.

It is shown that fructose accounts for the largest costs and amounts to 303.0 rubles (₽)/kg of polymer. When replacing fructose with glycerin, the cost of the carbon substrate is reduced by 41.6% when using purified glycerin and by 55% when using "raw" glycerin. When glucose is replaced with glycerin, the cost of the carbon substrate is reduced by 8.4% when using purified glycerol and by 29.4% when using "raw" glycerin. Thus, the technology for the synthesis of PHA on glycerol, developed, and implemented in a pilot version, provides an efficient process for producing polymers with a significant reduction in production costs.

TABLE 7.3   Costs of Carbon Substrates for the Production of 1 kg of PHA

Name of Substrate	Cost of Substrates P/ kg	Substrate Consumption, kg/kg PHA	Unit Cost of Substrate, P/kg PHA
Fructose	101.00	3.0	303.0
Glucose	69.00	2.8	193.2
Purified Glycerol "Duth glycerol refinery," Netherlands	59.00	3.0	177.0
Crude Glycerol "Prisma comercialexportadora de oleoquimicos LTDA,"Brazil	44.00	3.1	136.4

Source: Volova et al. [22].

7.3.2   COMPOSITES BASED ON PHA AND NATURAL MATERIALS

7.3.2.1   CHARACTERIZATION OF P(3HB) AND INITIAL NATURAL MATERIALS

Poly-3-hydroxybutyrate and natural materials (clay, peat, and birch wood flour) with considerably different properties (Table 7.4) were used to produce P(3HB)/filler blends: fine powder and paste prepared with ethanol (40% moisture content).

P(3HB) is a highly crystalline material, with the prevailing crystalline phase, and its degree of crystallinity (C_x) is usually above 60–65%; the C_x of the P(3HB) sample used in this study is 75%. XRD pattern of P(3HB) is showed, that diffraction peaks in 2θ = 13.4, 16.8, 20, 22.2 and 25.5°. XRD patterns of the fillers differed from the diffraction pattern of P(3HB). Wood flour showed distinct diffraction peaks at 2θ = 14.2, 19.6, and 25.5°, in contrast to the amorphized peat. X-ray examination of clay revealed numerous narrow diffraction peaks, and the most distinct one was at 23.5°. Clay had the highest C_x (53%). In wood flour and, especially, peat, amorphous phase prevailed, and their degrees of crystallinity were low: 26 and 9%, respectively.

Thermal properties of the polymer and fillers are different too. P(3HB) is a thermoplastic material, with a considerable difference between melting point (176°C) and thermal decomposition temperature (287°C). Its peak of thermal decomposition is in the 280°C. P(3HB) crystallization temperature is 108°C and melting enthalpy is 89.3 J/g. By contrast, none of the fillers is a thermoplastic material. Initial heating of these materials (and blends) gives a small peak at a temperature between 70 and 130°C, which is associated with evaporation of moisture and volatile organic compounds. After moisture evaporation, thermal decomposition of peat begins at 130°C and thermal decomposition of wood flour at 220°C. For clay, at a temperature between 60 and 200°C, moisture evaporates and decomposition and evaporation of volatile organic compounds begin (the process is endothermal); however, at temperatures between 200 and 400°C, decomposition is accompanied by heat release.

TABLE 7.4 Physicochemical Properties of Initial Materials and Blends

Sample	C_x,%	T_{melt}, °C	T_{cryst}, °C	T_{degr}, °C	Melting Enthalpy, (J/g)
Initial Materials:					
P(3HB)	75	176	108	287	89.3
Clay	53			200*	
Peat	9			130*	
Wood flour	26			220*	
Blends (P(3HB)/Filler = 70/30):					
P(3HB)/Clay:					
90:10	72	177	110	286	52.4
90:30	60	175	110	283	38.5
90:50	56	176	109	276	29.8
P(3HB)/Peat:					
90:10	60	174	107	288	64.1
90:30	48	176	108	292	59.7
90:50	42	177	110	290	48.7
P(3HB)/Wood Flour:					
90:10	65	178	107	279	66.9
90:30	47	175	106	273	51.3
90:50	44	171	104	274	45.4

* – start of $T_{degr.}$

Possible structural differences between the blends were detected using IR spectroscopy, which is employed to study the structures of various macromolecules. Variations in the intensity of the bands in the low-frequency range are indicators of both the crystalline to amorphous phase ratio in the blend and the type of interactions of the components in the blend.

FT-IR spectra of P(3HB) and the three fillers were taken in the 400–4000 cm^{-1}range. The spectra show not only the groups of carbon compounds (C=O, C-OH, CH$_2$, CH$_3$) contained in all initial materials but also the groups and compounds characteristic of individual components. These are CH, CH$_2$, CH$_3$, C-OH, and COOH groups for P(3HB), S-S-, C=C, C=N, and N-H groups for peat, Si-O-Si(AL) and Fe-O in α-Fe$_2$O$_3$ for clay, and -C-O-;-C-O-C-for birch wood flour. The IR absorption spectra of P(3HB) contain absorption bands corresponding to vibrations of the main structural components of polymers except the absorption bands of vibrations of the terminal C-OH and COOH groups. The bands of the ordered optical densities (crystalline phase) are in the 1228 cm^{-1} range while the bands of the amorphous phase are shifted to 1182 cm^{-1}. The spectra show distinct absorption bands of asymmetric stretching of CH$_3$- and CH$_2$-groups (2978 and 2994 cm^{-1}); symmetric stretching of CH- and CH$_2$ groups (2994 and 2934 cm^{-1}); stretching of conjugated (1687 cm^{-1}) and unconjugated (1720 cm^{-1}) carbonyl groups C=O; skeletal CH vibrations (599 cm^{-1}); and CH bending (622 cm^{-1}). PHAs are hydrophobic compounds, in which molecules of water interact with the hydrogen atom of the methyl group or with the oxygen atom of the carboxyl group through hydrogen bonding. The narrow bands at 3436 cm^{-1} are typical of the hydrogen bonded OH group. The 1687 cm^{-1} band is characteristic of hydrogen vibration in the OH group interacting with the oxygen atom through hydrogen bonding. The 2874, 2934, 2978, and 2994 cm^{-1} absorption bands are typical of the bonded OH group and may be a component of the COOH dimer group. In the 3000–2500 cm^{-1} range, there is a group of weak bands typical of the dimers of carboxylic acids.

In contrast to P(3HB) IR spectra, the IR spectra of peat contain absorption bands of the -S-S- bond groups (521 cm^{-1}); -C-O- groups characteristic of carbohydrates (1029 cm^{-1}); deformation vibrations of -CH$_2$- groups (718 cm^{-1}); skeletal vibrations of the benzene ring C=C (lignin), C=N, N-H (1513 and 1597 cm^{-1}), and stretching of the -OH groups at 3300 cm^{-1}. The IR spectra of clay contain absorption bands in the 776 and 795 cm^{-1} range, corresponding to stretching and symmetrical vibrations of Si-O-Si(Al) and Fe-O in α-Fe$_2$O$_3$; stretching vibrations of (AI-0), (Si-O-Al), and -Al$_2$O$_3$ in the 692 cm^{-1} range. In the 978 cm^{-1} range, there are absorption bands corresponding to (O-Si) deformation vibrations. The absorption frequency observed at 1635 cm^{-1} is indicative of deformation vibrations of H-O-H, which suggest water sorption on the surface of the mineral. The absorption band at 3615 and 3392 cm^{-1} corresponds to stretching and symmetric vibrations of AlO-H for clay bound water. IR spectra of birch wood flour show plane bending of the guaiacyl ring of lignin of leaves (1031 cm^{-1}), stretching of -C-O-;-C-O-C- (1235 cm^{-1}), skeletal vibrations of the syringyl ring of lignin of leaves (1324 cm^{-1}), stretching, and skeletal vibrations of benzene ring (1422 cm^{-1}, 1504 cm^{-1}, 1593 cm^{-1}), stretching of -C=O of conjugated and unconjugated groups (1652 cm^{-1}, 1734 cm^{-1}), and intramolecular and intermolecular stretching of -OH groups and water (3337–3342 cm^{-1}).

7.3.2.2   CHARACTERIZATION OF P(3HB)/FILLER BLENDS

Powdered P(3HB) and fillers were processed by cold pressing to produce pellets with a diameter of 6 mm and height 3 mm and granules with a diameter of 2 mm and length of between 2 and 4 mm (Figure 7.1).

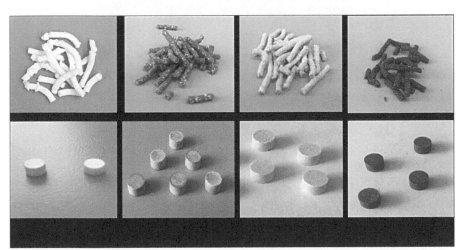

FIGURE 7.1   Granules and 3D forms obtained from PHA mixed with natural materials.
Source: T. Volova photo.

Pellets and granules prepared from blends with fillers that had different properties exhibited diverse surface structures (Figure 7.2). SEM images showed porous and rough surface microstructures. The surface of the

P(3HB)/wood flour pellets showed rectangular sawdust particles embedded in the surface layer; on the surface of the P(3HB)/peat pellets, there were asymmetrical heteraxial inclusions. On P(3HB)/clay pellets, clay inclusions were uniformly distributed over the surface.

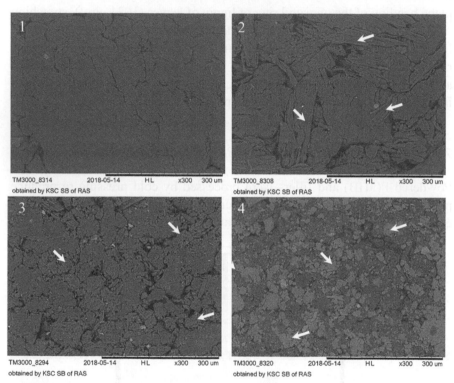

FIGURE 7.2 SEM images of pressed pellets (P(3HB)/filler = 70/30): 1-P(3HB); 2-P(3HB)/wood flour; 3-P(3HB)/peat; 4-P(3HB)/clay.

The blends were characterized using DSC, X-ray examination, and IR spectroscopy. IR spectrometry showed that characteristic absorption bands of initial components were separately retained in all blends: for P(3HB) those were CH, CH_2, CH_3, C-OH, and COOH groups; for peat-the additional and characteristic S-S-, C=C, C=N, N-H groups; for clay-Si-O-Si(AL) and Fe-O in α-Fe_2O_3; for birch wood flour: -C-O-;-C-O-C-. In the spectra of the blends, absorption bands were either separate bands or coinciding peaks, some of them transformed. An insignificant difference in numbers was within the limits of spectral error – 4 cm^{-1} (accumulation of 256 scans) and could result from increased moisture content of some of the samples. No new significant absorption bands were revealed in the blends, suggesting that no new chemical bonds were formed and that the blends prepared by the techniques employed in this study were physical mixtures.

Schematically, the process of obtaining mixtures from polymers and natural materials is presented in Figure 7.3.

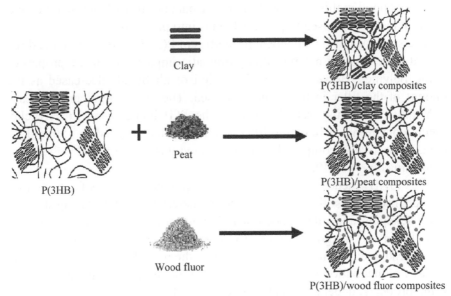

FIGURE 7.3 Schemes of initial materials and formulated mixtures (Sabu Thomas schemes).

Blending of P(3HB) with the fillers changed the properties of the polymer. The blends became progressively amorphous as the percentage of the fillers was increased (10, 30, and 50% fillings of the polymer with natural materials were tested). The degree of crystallinity of P(3HB) was decreased more noticeably in the blends with peat and wood flour (materials with low C_x). The C_x of the P(3HB)/peat and P(3HB)/wood flour blends with 30% filler decreased to 47 and 48%, respectively; the C_x of the blends with 50% of the same fillers decreased to 42 and 44%, respectively. That is, amorphous phase prevailed in these blends. In the P(3HB)/clay blends, the crystalline phase prevailed at all levels of filling of the polymer with clay; the C_x was no less than 56%. Changes in the crystallization kinetics of the polymer in the blends were associated with the rather high percentages of fillers (30 and 50%). This must have decreased the free volume necessary for nucleation of spherulites to occur, preventing the development of the crystalline phase and, thus, reducing it.

The study of thermal properties of the blends did not show any significant influence of the fillers on the temperatures of melting and thermal decomposition and the position of the T_{melt} and T_{degr} peaks in thermograms. However, the onset of thermal decomposition of the blends was shifted to the left relative to the initial P(3HB). The enthalpy of melting of the blends was significantly decreased (by a factor of 1.5–2.2). It is difficult to compare results obtained in this study with literature data because of data scarcity and different techniques used to process blends into products.

The study of physicochemical properties of P(3HB) blends with the three natural materials did not reveal any dramatic impairment of the properties of the polymer. However, Young's moduli of all blends decreased as the amount of the filler in the blend increased. The most significant decrease in Young's modulus was observed in the P(3HB)/peat and P(3HB)/clay blends containing 50% of the filler: the parameter dropped to 150 and 76 MPa, respectively, compared to Young's modulus of the initial polymer of approximately 360 MPa.

As P(3HB) blends can be considered as potential carriers for pesticides and fertilizers for soil application and as materials for fabricating degradable packaging and other goods, it is important to study their degradation behavior in soil. Degradation of P(3HB) blends with natural materials was first investigated in this study, showing that the mass loss of the blends over 35 days of incubation in soil varied between 30 and 50% of the initial mass of the products, depending on the type of the filler.

In this way, natural materials-clay, birch wood flour and peat-were used as fillers for degradable poly-3-hydroxybutyrate. The P(3HB)/filler to produce of pellets and granules. All initial materials and blends were investigated using IR spectroscopy, DSC, X-ray analysis, electron microscopy with microanalysis and X-ray spectrometry. Analysis of IR spectra showed that no chemical bonds were established between the polymer and fillers and that the blends were physical mixtures. Their temperature characteristics and degrees of crystallinity were lower than those of P(3HB), suggesting different crystallization kinetics of the blends. Degradation behavior of pellets and granules prepared from P(3HB)/filler blends in soil was similar to degradation behavior of pure polymer. The mass loss of all pellets and granules over 35 days of incubation in soil varied between 30 and 50% of the initial mass of the products, depending on the type of the filler. Thus, the blends prepared and tested in this study can be used to fabricate degradable packaging, greenhouse accessories, and delivery systems for fertilizers and pesticides.

7.3.3   A STUDY OF THE SUITABILITY OF MIXTURES OF P(3HB) AND NATURAL MATERIALS FOR THE CONSTRUCTION OF PLANT PROTECTION PRODUCTS OF A NEW GENERATION

The newest trend in research is development of new-generation slow-release pesticides, including fungicides, embedded in degradable matrices. The use of such formulations will decrease the amounts of pesticides applied to soil, ensure slow and targeted release of the active ingredients, and reduce the risk of uncontrolled distribution of the pesticides in the biosphere. The main condition for constructing such formulations is the availability of appropriate materials with the following properties: degradability and compatibility with global biosphere cycles, safety for living organisms and their nonliving environment, long-term presence in the natural environment and controlled degradation followed by formation of non-toxic products, chemical compatibility with pesticides, and processability by available methods.

Therefore, the purpose of this study was to construct and investigate slow-release fungicide formulations, with the active ingredient embedded in the degradable matrix of P(3HB) blended with readily available natural materials (peat, clay, wood flour) [23]. Slow-release fungicide formulations (azoxystrobin, epoxiconazole, and tebuconazole) shaped as pellets and granules in a matrix of biodegradable poly(3-hydroxybutyrate) and natural fillers (clay, wood flour, and peat) were constructed.

Antifungal activity of the matrix-loaded fungicides and the free fungicides is shown in Figure 7.4. The *in vitro* study showed that all formulations had a significant inhibitory effect on *F. verticillioides*: the size of the fungus colonies was smaller in the presence of the fungicides than in the control group. Photographs show *F. verticillioides* colonies on the nutrient medium in the presence of matrix-loaded fungicides and in the control group. The average radius of the fungus colonies in the Petri dishes with the experimental azoxystrobin formulations was comparable to the radius of the colonies in the dishes with pure fungicide (positive control) (1.3–1.7 cm) while in the dishes without any fungicide (negative control), the size of the fungus colonies was much larger (3.5–3.8 cm). Thus, the experimental formulations were as effective as the free fungicides. Similar results were obtained for the tebuconazole and epoxiconazole formulations. Formulations with different fillers produced similar fungicidal effects. No significant differences in the inhibitory effects on the growth of fungal colonies were found between the pellets and granules.

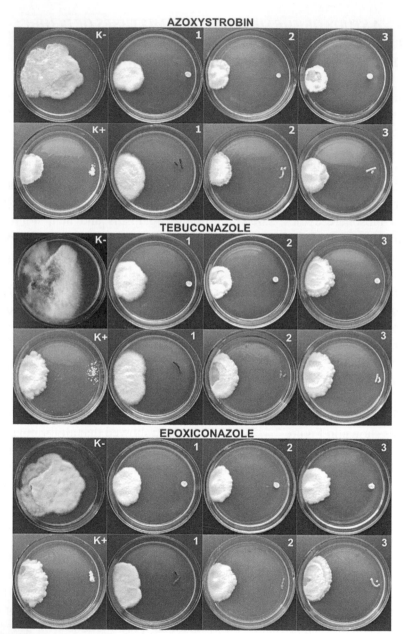

FIGURE 7.4 Sensitivity of *Fusarium verticillioides* to the fungicides in different forms: K⁻– negative control (without fungicide), K⁺ – positive control (active ingredient), 1 – pellets and granules of P(3HB)/peat/fungicide, 2 – pellets and granules of P(3HB)/clay/fungicide, 3 – pellets and granules of P(3HB)/wood flour/fungicide.

Source: Reprinted from Volova et al. [23].

Biodegradation of the experimental fungicide formulations was studied in laboratory soil microecosystems. Degradation behavior of the experimental formulations and the release of fungicides to soil were studied by determining the mass of the specimens during incubation of formulations in soil, taking into account residual concentrations of polymer and fungicides. These processes are certainly determined by various factors. During the incubation of the experimental granules and pellets in soil, only the polymer was degraded but the fillers (peat, clay, and wood flour) were not. Even if the formulations were degraded completely, the fillers would stay in the soil. Therefore, a study was performed to investigate degradation of poly(3-hydroxybutyrate) pellets and granules in soil. The granules were fully degraded in 126 days while degradation of pellets took longer-about 160 days. The type of the filler did not produce any major effect on degradation of the specimens, although specimens containing clay were degraded at a somewhat faster rate. Therefore, the residual mass of clay-containing specimens was the lowest – 55% of their starting mass.

All experimental formulations were capable of functioning in soil longer than the experimental period of 83 days, enabling gradual and slow delivery of fungicides to plants. Fungicide release from the polymer matrix was determined by measuring fungicide contents in the formulations before and after incubation of the specimens in soil. The active ingredients were released rather uniformly, especially from the pellets. By Day 20, a significantly higher amount of tebuconazole had been released from the granules than from the pellets. Fungicide concentrations in the experimental formulations with different fillers and fungicide concentrations in the soil were measured during incubation of the specimens in soil. Fungicide release kinetics was influenced by various factors such as kinetics of degradation of formulations in soil, shape of the specimens, composition of the matrix, solubility of the fungicides, and stability of the fungicides in soil. The fungicides accumulation in the soil released from the granules was somewhat higher than from pellets, especially in the early phase of incubation. The highest fungicide concentrations in the soil with the experimental granules were measured by the third week of the experiment. By contrast, the highest concentrations of the fungicides released from the experimental pellets were observed later, between week five and week seven of the experiment. The dynamics of accumulation of the three relatively poorly soluble fungicides were similar. Measurements of their residual concentrations in the specimens suggested that the fungicides were gradually released from both granules and pellets. After the fungicide concentrations in soil reached their maxima, no dramatic changes in this parameter occurred until the end of experiment (83 days).

The highest concentration of tebuconazole, whose solubility was higher than the solubility of azoxystrobin and epoxiconazole, varied between 20 and 26 µg/g soil, depending on the filler material (that is 20–26% of the fungicide content in the formulations before soil incubation) and the maximal concentrations of azoxystrobin and epoxiconazole in soil were 15 and 11 µg/g soil, respectively. These concentrations are corresponded to the recommended fungicide rates. The differences in their concentrations in soil were most likely caused by their dissimilar stabilities.

The studies have shown that all fungicide formulations had a pronounced inhibitory effect in vitro against the fungus *Fusarium verticillioides*. Biodegradation behavior of the experimental fungicide formulations in the soil was mainly determined by the shape of the specimens (granules or pellets) without significant influence of the filler type. The content of fungicides present in the soil due to degradation of the formulations and fungicide release was primarily determined by their solubility [23]. The use of P(3HB) and fillers as a matrix for embedding fungicides made it possible to develop slow-release formulations of fungicides, which can function in soil for a long time, enabling gradual and slow delivery of fungicides.

The next study was to develop ecofriendly herbicide formulations. Its main aim was to develop and investigate slow-release formulations of herbicides (metribuzin, tribenuron-methyl, and fenoxaprop-P-ethyl) of different structure, solubility, and specificity, which were loaded into a degradable matrix of poly-3-hydroxybutyrate (P(3HB)) blended with available natural materials (peat, clay, and wood flour) [24, 25]. Studies of biodegradation of the experimental herbicide formulations were performed in laboratory soil micro ecosystems. The specimens of herbicide loaded polymer/filler carriers were degraded gradually. The residual masses of the pellets and granules at the end of the experiment were 45–70% and 60–80%, respectively, of their starting masses. The composition of the matrix was a minor factor affecting mass loss of the slow-release herbicide formulations. Degradation kinetics of the majority of the experimental slow-release herbicide formulations was as follows: $\ln M_i = \ln M_0 + kt$, where M_0; M_i are the starting mass of the specimen and the mass of the specimen after degradation, g, k is degradation rate constant, and t is time, h. The validity of this equation for describing degradation kinetics of all the specimens was supported by the high accuracy of approximation (between 0.91 and 0.98). The highest values of degradation rate constant (0.00746 and 0.0094) were found for the specimens with metribuzin loaded into the clay- and peat-containing matrices and tribenuron loaded into the clay- and peat-containing matrices (0.0074 and 0.0077).

Thus, herbicide formulations were capable of functioning in soil longer than the experimental period of 83 day.

The effects of the experimental herbicide formulations on soil microflora were studied. After 83 days of incubation of the slow-release herbicide formulations in the soil, the total abundance of copiotrophic bacteria increased by a factor of 4–10 compared to the control, without herbicide application. The abundance of fungi in soil samples with slow-release metribuzin and tribenuron-methyl formulations increased by a factor of 1.3–3.6 on average while in the samples with the slow-release fenoxaprop-P-ethyl, the abundance of fungi decreased by a factor of 1.2–2.0. The effect of the herbicides on fungi was apparently determined by the type of the herbicide. Examination of the species composition of soil microflora at the end of the experiment (83 d) revealed certain changes in the proportions of the major taxa in soil samples with slow-release herbicide formulations compared to the control soil. *Bacillus* species, constituting up to 28.7%, prevailed in both groups of soil samples. The proportion of *Streptomyces* increased to 21–24.5% on average and *Pseudomonas* to 5.6–12.3%, while *Stenotrophomonas rhizophila* decreased to 8–9.7%. The type of the filler in formulations did not substantially influence the taxonomic composition of the dominant species in microbial communities. Thus, application of slow-release herbicide formulations did not decrease soil capacity for microbial degradation of complex organic compounds, including pollutants.

Herbicides released from the granules reached their peak concentrations in soil after 20 days of incubation while the herbicides released from most of the pellets reached their highest concentrations in soil much later, at Days 48–50. The solubility of the rapidly dissolving metribuzin was 1.2 mg/L, whereas the solubilities of the poorly soluble tribenuron-methyl and fenoxaprop-P-ethyl were 2 mg/L and 0.7 µg/L, respectively. Regardless of the type of the filler, metribuzin released from the pellets reached its peak concentrations in soil after 6–7 weeks of incubation: 60 to 80 mg/g soil, i.e., 50–80% of the loaded active ingredient, and remained steady for a long time, declining insignificantly by the end of the experiment. Residual metribuzin in the formulations was no more than 5–8% of its starting amount, i.e., the loaded herbicide was almost completely released to soil. Concentrations of the metribuzin released from the granules varied in a similar way, but the peak concentration was reached sooner: after three weeks for the specimens containing peat and after four weeks for the granules containing clay and wood flour. Metribuzin release from the pellets corresponded to the Fickian diffusion mechanism and from the granules to the non-Fickian transport, regardless of the filler type, suggesting considerable structural differences between the carriers. Thus, metribuzin was steadily and gradually released from both pellets and granules and, hence, was present in the soil over extended time periods.

So, herbicide formulations enabling slow controlled release of the active ingredients were fabricated. The degree of solubility of the herbicide considerably affected its release rate and degradation of the formulations in soil: the rapidly dissolving metribuzin and tribenuron-methyl were released at faster rates than the less readily soluble fenoxaprop-P-ethyl. Loading of the herbicides into the polymer matrix composed of the slowly degraded P(3HB) and available natural materials enabled both sustained function of the formulations in soil (lasting between one and a half and three or more months) and stable activity of the otherwise rapidly inactivated herbicides such as tribenuron-methyl and fenoxaprop-P-ethyl.

7.4 CONCLUSION

Two approaches have been investigated in order to reduce the cost and increase the availability of the use of biodegradable polymers of microbiological origin (polyhydroxyalkanoates, PHA):(1) expanding the raw material base and attracting glycerol as a carbon substrate (large-tonnage waste from the production of biodiesel); and (2) filling the polymer with natural materials for industrial use.

An effective technology for the synthesis of destructible polymers was developed using a new and more affordable substrate compared to sugars-glycerol. The *Cupriavidus eutrophus* B-10646 strain was studied as a product of polymers on glycerin, and the kinetic and production characteristics of the strain providing polymer yields on glycerol of various cheat, comparable with the process on sugars, were studied. Replacing sugars with glycerol provides a reduction in the cost of a carbon substrate in the production of PHA.

Mixtures were formed using P (3HB) and available natural materials (clay, peat, birch sawdust). From mixtures in the form of dry powders and raw paste, mixed forms in the form of granules and extruded 3D forms were obtained; their structure, physicochemical, and mechanical properties were studied. A comparative study of the influence of the type of form and method of manufacture (by pressing powders, extrusion from melts, mortar technologies) on the intensity of biodegradation and the duration of functioning in the soil substantiated the feasibility of using forms that can provide plants with preparations during the growing season (from 1.5 to 3.0 and more months) of two methods-obtaining granules from crude paste of polymer and filler mixtures and cold pressing of powders. Based on the results obtained and their analysis, a pioneer family of prolonged preparations of herbicidal and fungicidal action for pre-emergence soil application was constructed.

Regardless of the geometry of the developed formulations, the composition of the mixed base, the solubility of pesticides in water and their stability, it is shown that the achieved concentrations of active substances in the soil are within the limits corresponding to the recommended application rates. The results showed the suitability of the developed and studied mixtures and forms of them for the construction of long-term new generation plant protection products.

ACKNOWLEDGMENT

This study was financially supported Project "Agropreparations of the new generation: a strategy of construction and realization" (No 074-02-2018-328) 2018–2019 in accordance with Resolution No 220 of the Government of the Russian Federation of April 09, 2010, "On measures designed to attract leading scientists to the Russian institutions of higher learning."

KEYWORDS

- differential scanning calorimetry
- free fatty acids
- pilot production
- polyhydroxyalkanoates
- x-ray diffraction

REFERENCES

1. Urbanek, A. K., Rymowicz, W., & Mirończuk, A. M., (2018). Degradation of plastics and plastic-degrading bacteria in cold marine habitats. *Appl. Microbiol. Biot., 102*, 7669–7678. https://doi.org/10.1007/s00253-018-9195-y (accessed on 3 August 2020).

2. Qaiss, A., Bouhf, R., & Essabir, H., (2015). Characterization and use of coir, almond, apricot, argan, shell, and wood as reinforcement in the polymeric matrix in order to valorize these products, In: Hakeem, K., Jawaid, M., & Alothman, O. Y., (eds.), *Agricultural Biomass Based Potential Materials* (pp. 305–339). Cham: Springer,Switzerland. https://doi.org/10.1007/978-3-319-13847-3_15 (accessed on 3 August 2020).

3. Chen, G. Q., (2010). Plastics from bacteria: Natural functions and applications. *Microbiology Monographs*. Springer-Verlag, Berlin.

4. Chen, G. Q., (2012). New challenges and opportunities for industrial biotechnology. *Microb. Cell Fact., 11*, 111. doi: 10.1186/1475-859-11-111.

5. Volova, T., Vinnik, Y., Shishatskaya, E., Markelova, N., & Zaikov, G., (2017). *Natural-Based Polymers for Biomedical Applications*. Apple Acad. Press, Canada.

6. Volova, T., Shishatskaya, E., Prudnikova, S., Zhila, N., & Boyandin, A., (2019a). *New Generation Formulations of Agrochemicals: Current Trends and Future Priorities* (p. 286). Toronto-Canada: CRC/Taylor & Francis: Appl. Acad. Press. ISBN: 9781771887496.

7. Koller, M., Maršálek, L., De Sousa, D. M. M., & Braunegg, G., (2017). Producing microbial polyhydroxyalkanoate (PHA) biopolyesters in a sustainable manner. *New Biotechnol., 37*, 24–38. https://doi.org/10.1016/j.nbt.2016.05.001 (accessed on 3 August 2020).

8. Hermann-Krauss, C., Koller, M., Muhr, A., Fasl, H., Stelzer, F., & Braunegg, G., (2013). Archaeal production of polyhydroxyalkanoates (PHA) co-and terpolyesters from biodiesel industry-derived by-products. *Archaea* (p. 10). Article ID 129268. doi: 10.1155/2013/129268.

9. Kourmentza, C., Plácido, J., Venetsaneas, N., Burniol-Figols, A., Varrone, C., & Gavala, H. N., (2017). Reis MAM recent advances and challenges towards sustainable polyhydroxyalkanoate (PHA) production. *Bioengineering, 4*(2), e55. doi: 10.3390/bioengineering4020055.

10. Torres-Giner, S., Montanes, N., Boronat, T., Quiles-Carrillo, L., & Balart, R., (2016). Melt grafting of sepiolite nanoclay onto poly(3-hydroxybutyrate-co-4-hydroxybutyrate) by reactive extrusion with multi-functional epoxy-based styrene-acrylic oligomer. *Eur. Polym. J., 84*, 693–707.https://doi.org/10.1016/j.eurpolymj.2016.09.057 (accessed on 3 August 2020).

11. Gunning, M., Geever, L., Killion, J., Lyons, J., & Higginbotha, C., (2013). Mechanical and biodegradation performance of short natural fiber polyhydroxybutyrate composites. *Polym. Testing, 32*, 1603–1611. https://doi.org/10.1016/j.polymertesting.2013.10.011 (accessed on 3 August 2020).

12. Angelini, S., Cerruti, P., Immirzi, B., Scarinzi, G., & Malinconico, M., (2016). Acid-insoluble lignin and holocellulose from a lignocellulosic biowaste: Bio-fillers in poly(3-hydroxybutyrate). *Eur. Polym. J., 76*, 63–76.https://doi.org/10.1016/j.eurpolymj.2016.01.024 (accessed on 3 August 2020).

13. Khunthongkaew, P., Murugan, P., Sudesh, K., & Iewkittayakorn, J., (2018). Biosynthesis of polyhydroxyalkanoates using Cupriavidus necator H16 and its application for particleboard production. *Journal of Polymer Research, 25*, 131. https://doi.org/10.1007/s10965-018-1521-7 (accessed on 3 August 2020).

14. Luigi-Jules, V., Clement, M. C., Alan, W., Des, R., Bronwyn, L., & Steven, P., (2018). Wood-PHA composites: Mapping opportunities. *Polymers, 10*,751–766. https://doi.org/10.3390/polym10070751 (accessed on 3 August 2020).

15. Du, C., Sabirova, J., Soetaert, W., & Lin, S. K. C., (2012). Polyhydroxyalkanoatesproduction from low-cost sustainable raw materials. *Curr. Chem. Biol., 6*, 14–25. doi: 10.2174/2212796811206010014.
16. Możejko-Ciesielska, J., & Kiewisz, R., (2016). Bacterial polyhydroxyalkanoates: Still fabulous. *Microbiol. Res., 192*, 271–282. doi: 10.1016/j.micres.2016.07.010.
17. Sabbagh, F., & Muhamad, I., (2017). Production of polyhydroxyalkanoates as secondary metabolite with main focus on sustainable energy. *Renew. Sust. Energ. Rev., 72*, 95–104. https://doi.org/10.1016/j.rser.2016.11.012 (accessed on 3 August 2020).
18. Fernández-Dacosta, C., Posada, J., Kleerebezem, R., Cuellar, M., & Ramirez, A., (2015). Microbial community-based polyhydroxyalkanoates (PHAs) production from wastewater: Techno-economic analysis and ex-ante environmental assessment. *Bioresour. Technol., 185*, 368–377. doi: 10.1016/j.biortech.2015.03.025.
19. GHG emission reductions from world biofuel production and use (2015). *Global Renewable Fuels Alliance.* https://www.epure.org/media/1298/final_report_ghg_ emissions_biofuels_2015-3.pdf (accessed on 14 August 2020).
20. Statista, (2018). *Leading Biodiesel Producers Worldwide in 2017.* By country (in billion liters). https://www.statista.com/statistics/271472/biodiesel-production-in-selected-countries (accessed on 3 August 2020).
21. Volova, T., Demidenko, A., Kiselev, E., Baranovskii, S., Shishatskaya, E., & Zhila, N., (2019b). Polyhydroxyalkanoate synthesis based on glycerol and implementation of the process under conditions of pilot production. *Applied Microbiology and Biotechnology, 103*(1), 225–237 https://doi.org/10.1007/s00253-018-9460-0 (accessed on 3 August 2020).
22. Volova, T., Kiselev, E., Zhila, N., & Shishatskaaya, E., (2019c). Synthesis of PHAs by hydrogen bacteria in a pilot production process. *Biomacromol., 20*, 3261–3270. doi: 10.1021/acs.biomac.9b00295.
23. Volova, T., Prudnikova, S., Boyandin, A., Zhila, N., Kiselev, E., Shumilova, A., Baranovsky, S., et al., (2019d). Constructing slow-release fungicide formulations based on poly-3-hydroxybutyrate and natural materials as a degradable matrix. *Journal of Agricultural and Food Chemistry, 67*, 9220–9231. https://doi.org/10.1021/acs. jafc.9b01634 (accessed on 3 August 2020).
24. Kiselev, E. G., Boyandin, A. N., Zhila, N. O., Prudnikova, S. V., Shumilova, A. A., Baranovskiy, S. V., Shishatskaya, E. I., Thomas, S., & Volova T. G., (2020). Constructing sustained-release herbicide formulations based on poly-3-hydroxybutyrate and natural materials as a degradable matrix. *Pest Manag Sci. 76*, 1772–1785. https:// doi.org/10.1002/ps.5702.
25. Thomas, S., Baranovsky, S., Vasiliev, A., Kiselev, E., Kuzmin, A., Nemtsev, I., Sukovatyi, A., et al., (2019f). Thermal, mechanical, and biodegradation studies of biofiller based poly-3-hydroxybutyrate biocomposites. *International Journal of Biological Macromolecules.*

CHAPTER 8

Membrane Technology for Green Engineering

SUPRIYA DHUME and YOGESH CHENDAKE

Department of Chemical Engineering, Bharati Vidyapeeth (Deemed to be) University, College of Engineering, Pune, Maharashtra, India, E-mail: yjchendake@bvucoep.edu.in (Y. Chendake)

ABSTRACT

Green technology is that to operate ecological skills for the development and application of merchandise, instrumentality, and systems to conserve the natural resources and surroundings. This field is moderately innovative. It has developed rapidly as people have become aware of the detrimental effects of environmental variation including global warming and adverse effects on the living beings. It is believed that growth in green technology will lead to world-wide, sustainable, and economic powers that impact economics, societies, and way of life in the future. Application of green technology has multiple challenges while it can provide many opportunities during real-life chemical processes. The membrane technology can provide an economical pathway to many of these challenges. Here applicability of membranes during green technology is discussed. Over past few years, membrane separation process has become one of the major separation technologies. It is rapidly becoming important part of green technology due some of its benefits. In membrane processes does not require addition of chemicals, it works on physical separation. This improves the feasibility of recycling the recovered chemicals or using it in further processes. Additionally, membrane technology has relatively low energy requirements, easy to design and integrated nature may advance, and can be integrated with current processes. This gives benefits of reduced production cost, equipment size, energy consumption and waste generation and improving controls and process flexibility [1–3].

The works would consider two aspects of membrane technologies for process enhancements; as process integration and effluent treatment. Process integration would be focused on for improvements in economical separation, recovery, and recycle; this would reduce generation of waste and effluent component generation, thus enhancing process financial system. Membrane separation technologies are very helpful in effluent treatment. It is investigated for reduction in effluent volume and its treatment for components to recover. The work starts by presenting an overview of membrane operations. It would be followed for application of membrane processes for process improvement and *in situ* recovery of components. This would be followed by the effluent treatment aspects. Such recovery of components harmful to environment and their further utilization by membrane processes would increase the process economy and product purity. To get superior removal of contaminants compared to conventional technologies, the membrane technology would provide better support to system. The conventional technologies frequently required continuous modification of conditions and use of chemicals. This means a high level of operator involvement and the risk of contaminants in the effluent [4–7].

8.1   INTRODUCTION

8.1.1   GREEN TECHNOLOGY

Green Technology is nothing but modification of process parameters for the expansion and application of products, equipment, and systems to conserve the natural environment and resources. In case of unavoidable effluent and waste generation, it requires it recovery and recycle to system for improvement in conversion and yield for desired product, with reduction or elimination of hazardous effluent generation. It helps to reduce the volume of effluent generation; avoid harmful effluent stockpiling or contamination with environment. It shows the harmful impact on human activities. Green technology refers to use or synthesis of products and equipments which satisfy the following criteria:

1.   It reduces the ecological degradation;
2.   It has low greenhouse gas emission;
3.   It is safe and provide healthy and good environment;
4.   Conservation of natural resources and energy usage;
5.   It assists to maximize consumption of renewable resources, than diminishing mineral and petroleum oil ones.

Overall, it can be said that, green technology is an eco-friendly clean technology contributes to sustainable development to conserve the natural resources and environment. Development of such synthesis route, processes, equipments, and systems will meet the demands of present and future generations. These developments would help humans, to eliminate the utilization or generation of hazardous substances [8–12].

The principles of green technology are relevant to chemical products including its usage, design of process, manufacturing, and final disposal. Green technology is also referred to be sustainable developments. This technology development and chemical engineering focused on such designed process which reduces or eliminates the generation of hazardous substances. The environmental engineering focuses on the effects of polluting chemicals on nature, where green technology focuses on the environmental impact of chemicals their synthesis and waste reduction. It consists of industrial approaches to prevent the pollution and reducing consumption of non-renewable resources, enhancing output within defined resources, optimal conversion, and reduction in energy consumption.

Green technology concerns are also related to production of food and its process which requires considerable usage of land. Such land-use changes and shows an incredible reduction in biological assortment, aquatic organic process with phosphorus substances which caused by over-fertilization, water shortages due to irrigation, climate changes (CCs), eco-toxicity, human effects of pesticides, etc. They are lead to contamination of surroundings with different toxic substances contains arsenic, cadmium, chromium, lead, and mercury. These are the priority metals that are having high level of toxicity. This toxicity depends upon several factors including their dose, route of exposure, and chemical species [13–15].

Another one of the important aspects of chemical processing is energy conversion. Some of the energy sources considered today is fossil fuels, hydro energy, other forms of electrical energy generation (tidal, wind, atomic, etc.). All such conventional energy sources affect the environment differently. Fossil fuels like coal, oil, and natural gas are considerably more harmful than renewable energy sources. Frequently its usage leads to air and water pollution. This intern results in damage of public health, wildlife, and habitat loss, water resource contamination, land use, and global warming emissions. This can be overcome by utilization of renewable energy resources. These renewable energy resources would replace fossil fuels, thereby reducing emissions that cause CC. Increasing renewable energy allows decreasing the costs described, and that is obviously very economically beneficial in combination with the environmental benefits. There are many more ways green energy benefits the economy, directly, and indirectly. By reducing the use of traditional energy sources, natural ecosystems can be maintained to the best conditions. Further use of Green Energy can be economically attractive. The economic benefits of green energy can be summarized as follows:

1. Increase in energy efficiency.
2. Lower energy cost from the renewable sources viz., solar, wind, and biomass.
3. Nowadays, carbon emission is being reduced and same has been mandated through various laws. Some of the laws have established laws related to carbon credits and fines for higher release of these gases. Today fuel cost has to be finalized considering, carbon emission.

To achieve these goals of green technology, there is a need for design of chemical processes and its development or application of products; equipments for conservation of natural resources are beneficial to environment. In the environmental point of view, the necessity is to avoid use or generation of hazardous substances to reduce the pollution in environmental point of view. And the process modification and propose feature must think to reduce or eliminate energy and avail the economical benefits.

8.1.2   NEED AND GOALS

As discussed until now, Green Technology possesses multiple benefits. They are achieved by following principals of green technology:

- **Prevent Waste:** Chemical production is design such a way that remains no waste to treat or clean up.
- **Atom Economy:** Design of synthesis in such a way to the final product contain highest amount of initial material. Waste a small number of or no atoms.
- **Design Less Hazardous Chemical Production:** Design synthesis process to generate the substances with minimum or no toxics to humans or environment.
- **Design Harmless Chemicals and Products:** The chemical products are designing in such a way that has little or no toxicity.
- **Use Unhazardous Solvents and Reaction Conditions:** Avoid using hazardous solvents, separation agents, or other auxiliary chemicals. If necessity to use these chemicals, use safer ones.

- **Increase Energy Efficiency:** Run chemical reactions at room temperature and pressure whenever possible.
- **Use Renewable Feedstocks:** Use of raw materials which are renewable. The source of renewable feedstocks is frequently the agricultural products or the wastes of other processes; the source of depletable feedstocks is often fossil fuels (petroleum, natural gas, or coal) or mining operations.
- **Avoid Chemical Derivatives:** Avoid using blocking or defensive groups or any temporary modifications if possible. Derivatives use additional reagents and generate waste.
- **Use Catalysts, Not Reagents:** Minimize waste by using catalytic reactions. Catalysts are efficient with small amount and can carry out a single reaction many times. They are preferable to stoichiometric reagents, which are used in excess and carry out a reaction only once.
- **Design Chemicals and Products to Degrade After Use:** Design chemical products to break down to harmless substances after use so that they do not accumulate in the environment.
- **Analyze in Real-Time to Prevent Pollution:** Include in-process, real-time observation, and control during synthesis to diminish or eliminate the development of byproducts.
- **Minimize the Potential for Adversity:** Design chemicals and their physical forms (solid, liquid, or gas) to diminish the potential for chemical accidents together with explosions, fires, and releases to the environment.

Hence there is a need to improve or modify the process so as to incorporate the Green Technology principals. At many places, economical separation of components and their composition variations can be useful to achieve the desired progress towards Green Technology. Membrane processes and their process applications are an excellent way for the same. Membrane-based systems carry out physical separations based upon component and membrane properties. Additionally, the separation conditions can be tuned as per the need. Further section deals with membrane processes and their applications towards Green Technology.

8.2   MEMBRANE PROCESSES

Membrane is nothing but semipermeable barrier, which allows selective transport of component across the same. The transport and separations are in the same phase. This reduces requirement of heat separation of components. Components are separated upon their molecular size, shape, charge, interactions, etc. with use of pressure and particularly considered semi-permeable membranes. This separation of components does not require accumulation of chemicals or any contaminations. The separation is carried out in single-stage without phase inversion hence generation of hazardous waste is eliminated.

Membrane separation in green technology is describe in such a way that, with help of membrane process reduction in byproduct/waste due to less usage of chemicals during production. It can be combined with the synthesis processes for continuous separation which gives the benefits of recovery and reuse of waste/byproduct can be usable as raw feed to the system. This reduces generation of waste, and effective recycle enhances conversion along with conservation of natural resources.

Membrane separation technology is physical separation. It provide excellent alternative for recovery of components in pure form. This helps in the waste processing and recovery of components. These recovered components can be recycled to the processing streams. A proper and careful design of membrane properties and separation parameters can be used to recover or separate the components from effluent and utilize economically. It would reduce or eliminate the generation of hazardous substances in environment. These hazardous substances are harmful to the nature/environment.

Contamination of these hazardous components can affect CC, availability of harmless and sufficient water supply, and food production. Additionally, it would require additional energy for the treatment, which in turn would affect the emission of hazardous substances in the atmosphere. This would have a vital impact over the water cycle, sterilization precipitation patterns, and poignant the supply and quality of surface and groundwater, agricultural production, and associated ecosystems. Water usages for agricultural is up to the seventieth each primary food production and food process is critically dependent upon reliable installation and adequate water quality. Each variety of activities may also have serious effects on water resources. This would affect human and surrounding to a greater extent. Recovery and removal of such components before their discharge into the atmosphere can prove highly beneficial to fight these demons. The membrane-based processes can provide answers to these issues.

As mentioned up till currently, there is have to be compelled to minimize losses in terms of waste generated through reaction selection, method style and use or recycling. This may cause effective use of obtainable resources and generate the littlest doable quantity of stuff. It might give massive economical and ecological advantages to the economic makers. Use of membrane methods will be one among the wonderful choices throughout process modifications. The membrane processes of microfiltration (MF), ultra-filtration, and reverse osmosis (RO) supply glorious potential for continuous removal of elements from effluent streams. Choice of the optimum method could be a operate of material properties and physical forms. Membrane separation technologies cannot build the contaminants disappear; but, they are extraordinarily effective at concentrating the contaminants, i.e., a pretreatment part, recovery of part for his or her applications in any processes. The membrane method has a plan of use or convalescent specific, i.e., parts for re-use through the applying of membrane separation technologies.

Due to increasing limitations placed on waste and effluent disposal, innovative ways of dangerous waste treatment are required; which can be fulfilled by membrane technologies. The benefit of these technologies is that they can be tuned as per requirement and application in hand. They can be combined with current process to enhance the output, decreases the byproduct formation and scale back or eliminate waste generation. In recent years, membrane separation has gained increasing quality for these applications. The three major varieties of membrane separation technologies in use nowadays square measure reverse diffusion, immoderate filtration, and electrochemical analysis.

The popularity and applicability of membrane-based processes in increasing due to subsequent reasons. The combination of membranes based separation system can be done in line with the synthetic streams. This helps to monitor and control the product properties and desired environment.

A strong increase within the world's population and an eternal growth of commercial productivity the global energy consumption continues to extend. On the other hand, the fossil energy resources viz. oil, gas, and coal are progressively exhausted. So as to secure the energy offer for the close to future energy-saving measures ought to be taken and also the development of energy-saving technologies ought to be aroused. During this respect membrane technology is an associate emerging technology with several prospects as associate energy-saving separation technology in a range of classical separation techniques like distillation, evaporation, refrigeration, condensation, a phase change occurs and the energy consumption is relatively high. Membrane technology is one amongst the new separation techniques with comparatively low energy consumption since no phase change takes place.

In membrane process continuous separation is, occurs though some unwanted components are clogging the membrane to avoiding this unnecessary derivatization a little modification has to done on temporary basis. Like doing some changes in membrane, preparation helps to reduce the clogging or give backwash to the system to avoiding the clogging. Also in life processing and control helps to avoid hazardous products. Moderate separation conditions and minimize potential of chemical accidents like releases, explosions, fires, and formation of hazardous components and effluents. Further benefits of membrane processes can be summarized as below.

8.3 BENEFITS OF MEMBRANE SEPARATIONS

Membrane separations are often used for very large number of separation because membrane processes separate at the molecular scale up to a scale at that particle can be seen. Membranes have intensely high property for the separation of components. There can be a big struggle of management over separation property by having larger range of polymers and inorganic media usage as membranes. Energy necessities area unit sometimes low because membrane separations typically don't need a phase transition. They sometimes supply a simple and straightforward operation and low maintenance method possibility as they are doing not have moving parts, complicated management schemes, and little subsidiary instrumentation. Membranes will have intensely high property for the separation of parts. Membrane processes area unit sometimes higher for the surroundings once there is a usage of comparatively straightforward and non-harmful materials and they will recover minor however valuable constituents from a mainstream with a minor energy value. Due to these benefits, many membrane processes have been designed and utilized in green technology.

8.4 MEMBRANE PROCESSES IN GREEN TECHNOLOGY

Membrane filtration could be a technique of separating particles in liquid solutions or gas mixtures. This system is employed in a very big selection of applications starting from dairy process to waste treatment. The four main ways of pressure-driven membrane filtration are MF, Ultrafiltration, Nanofiltration, and RO, so as of decreasing pore size. Their brief properties are summarized below:

1. **Microfiltration (MF):** These membranes have pore sizes ranging from 0.1 to 10 μm. They are mainly used for the removal of huge particles, colloids, and bacteria from feed streams. This is often

particularly well-liked within the food trade for treating waste material before discharging it to a municipal sewer.

2. **Ultra-Filtration (UF):** It is a process that is similar to MF, but with smaller pore size, ranging from 0.01 to 0.1 μm. UF membranes are used in removal of viruses and polypeptides, and are extensively used in protein concentration and wastewater treatments.

3. **Nanofiltration (NF):** These membranes are similar to RO membranes in that they contain a thin-film composite layer (<1 μm) on top of a porous layer (50 to 150 μm) for small ion selectivity. NF membranes are rejecting multivalent salts and uncharged solutes, whereas permitting some monovalent salts to accept. They operate at lower pressures than RO membranes, creating them ideal for achieving the associate best combination of flux and rejection.

4. **Reverse Osmosis (RO):** These membranes are even tighter than NF membranes, and are ready to reject all monovalent ions whereas permitting water molecules to meet up with in aqueous solutions. They additionally take away viruses and microorganisms found in feed solutions. Common applications for RO filtration embrace seawater desalination and industrial water treatment.

The other main methods of concentration driven membrane filtration are pervaporation (PV), gas separation, dialysis, and electrodialysis. Chemical analysis could be a method during which the solute, usually an electrolyte, transfers across the membrane driven by the distinction in concentration between the two sides of the membrane. Two conditions ought to be consummated for the method to be effective:

i.   The concentration on the permeate aspect has got to be unbroken low so the driving force remains as high as possible; and

ii. The pressure should be low, and stay low throughout the method, so a counter flux of solvent does not end in feed dilution.

8.5   ELECTRODIALYSIS

Electrodialysis could be a method within which ion-selective membranes measure used alongside an electrical field normal to the membrane phases. An electrodialysis stack consists of parallel compartments separated alternately by cation-exchange and anion-exchange membranes. The feed is introduced into every compartment, so power-assisted by the electrical field and selected by the membranes, the cations, and also the anions are transferred in opposite directions to the neighboring compartments. Betting on the sort of membrane used, ions are often selected consistent with their valence. So this method is often used for fractionating ions of various valences. The most application of electrodialysis is in demineralization normally and within the chemical process of saltwater.

Membrane electrolysis could be a method whereby each electrode reactions, i.e., the cathodic reduction additionally because the electrode oxidization, are coupled to the transport and transfer of charged ions. In membrane electrolysis, the electrode reaction is crucial to the particular separation method. The aim of the membrane is to separate the anode loop (anolyte) from the cathode loop (catholyte) by a fluid, so as to avoid unwanted secondary reactions, thus on mix the electrode reaction with a separation step or to isolate individually the products formed on the electrode. In water electrolysis, such products is also in an exceedingly gaseous form such as oxygen and hydrogen element additionally because the acids (H+) and bases (OH−) fashioned on the electrode or the mixture of gas halogen and hydroxide answer and element as in common sodium chloride electrolysis.

8.6   PERVAPORATION (PV)

Pervaporation (PV) is that the combination of the selective separation and transfer of an element across the membrane and its evaporation on the permeate aspect. So as to attain evaporation the pressure on the permeate aspect should be specified the partial pressure of this element is not up to its saturation vapor pressure. Therefore, although the driving force for transfer is the difference in activity of the transferred species, this is the result of applying a vacuum on the permeate side.

PV is dearer than different membrane processes due to the specified offer of warmth to provide the evaporation. It is, therefore, utilized in the separation of parts from mixtures that area unit tough to treat, like azeotopic solutions and chemical compound mixtures, that standard processes would be additional pricey.

8.7   MEMBRANE DISTILLATION (MD)

Membrane distillation (MD) could be a separation method wherever a microporous hydrophobic membrane separates two liquid solutions at totally different temperatures. The property of the membrane prevents mass transfer of the liquid, whereby a gas-liquid interface is made. The gradient on the membrane leads to a vapor pressure distinction, whereby volatile elements within the offer combine evaporate through the pores (10 nm–1 μm) and, via diffusion and/or convection of the compartment with high vapor pressure, area unit transported to the compartment with low vapor pressure wherever they are condensed within the cold liquid/vapor phase. To offer solutions that solely contain non-volatile substances, like salts, water vapor are going to be transported through the membrane whereby dematerialized water is obtained on the distillation-side and an extra targeted salt flow on the supply-side.

8.8   PROCESS INTENSIFICATION (PI)

One of the essential challenges presently facing the planet is "to support sustainable industrial growth." A doable answer is obtainable by process intensification (PI), a design approach presenting existing edges in producing and process. Extra edges may be summarized as shrinking equipment size, boosting plant effectiveness, saving energy, reducing capital costs, increasing safety, minimizing environmental impact, and maximizing raw materials exploitation.

Membrane methods deal with the goals of PI; as a result of they need the potential to interchange conventional energy-intensive techniques, to accomplish the selective and economical transport of specific elements, and to enhance the performance of reactive processes. On a variety of occasions, industrial separation processes in trade were replaced or power-assisted by membrane separation processes with important reductions in cost, energy, and environmental impact.

It can be seen from the following examples that the use of membranes for continuous processing and process improvements can be useful to achieve the goals for PI. The membrane engineering contributes to the comprehension of the principles of green technology. Continuous separation and recovery of components, separation in milder conditions, recycle of separated components, improve reaction kinetics, conversions, energy conservation, energy generation at higher efficiency, and improved product quality are all the benefits of membrane technology. They help to reduce cost of production and provide an environmental-friendly system. Let go to the examples for their complete descriptions. It can be seen from the following examples, viz., membrane reactor for esterification, fermentation-based component synthesis, ethanol concentration, and dairy processing.

8.8.1   MEMBRANE REACTOR-ESTERIFICATION REACTION

Esterification of acids and alcohols could be a typical example of an equilibrium restricted reaction that generates water as a by-product. The conversion is mostly low due to restrictions forced by thermodynamic equilibrium. To urge a high organic compound yield, it is necessary to shift the position of the equilibrium to the organic compound facet by either employing a massive way over one among the reactants (usually alcohol).

One of the foremost ordinarily investigated samples of esterification could be a chemical process that happens between carboxylic acid and alcohol to get esters. The reaction takes place in acidic environments. During this method, water is additionally generated as byproduct.

The reaction between carboxylic acid and alcohol is understood because the Fischer esterification. It is one amongst the foremost vital reactions of carboxylic acids. Most carboxylic acids are appropriate for the reaction; however, the alcohol ought to usually be a primary or secondary alkyl. Tertiary alcohols are liable to elimination. Once an acid and an alcohol are mixed along, no reaction takes place. However, upon addition of chemical action amounts of an acid, the two components are mix with equilibrium method to contribute an ester and water. The presence of the acid catalyst within the mechanism of ester formation helps in two ways: It causes the carbonyl perform (makes the carbonyl carbon a lot of electrophilic) to bear nucleophilic attack by the alcohol; and protonation of the hydroxyl group offers water, that could be a superior leaving group (i.e., weaker base) within the elimination step. Ordinarily used catalysts for a Fischer esterification sulfuric acid, p-toluenesulfonic acid, and Lewis acids.

Esterification reactions are generally reversible processes and generate water as part of production of ester. The example of acetic acid and methanol reaction at 60°C is shown these phenomena. This reaction shows an investigation into the impact of water on liquid phase sulfuric acid. In order to moderate, the effect of water on the catalysis as a result in a reversible reaction, initial reaction kinetics were measured using a low concentration of sulfuric acid and different initial water concentrations. It was found that the catalytic activity of sulfuric acid was strongly inhibited by water. The catalysts lost up to 90% activity as the amount of water present increased. The decreased activity of the catalytic protons is suggested to be caused by preferential solvation of them by water over methanol. The results indicate that, as esterification progresses and byproduct water is produced, deactivation of the sulfuric acid catalyst occurs.

As explain before esterification is an equilibrium reaction and water is one of the products of this reaction. The generalized esterification reaction is already discussed above.

In any esterification, reaction water is the byproduct of the reaction. If this water is not removed immediately after the formation of esters then it reacts to get back the reactants due to reversible reaction conditions. Therefore, as per the thermodynamic considerations, to produce more ester and shift the equilibrium in the forward direction; water should be removed immediately.

The membrane reactors are working very effectively for continuous removal of water generated during process. Applications of membrane reactors for water removal during catalytic reactions in food, pharmaceutical, cosmetics, and petrochemical industries are well investigated and established. Currently, global target in this direction is design of a compact and more efficient catalytic membrane reactor. By applying water removal, two objectives have been pursued:(i) overcoming the thermodynamic limitations of the reaction; and (ii) avoiding catalyst poisoning.

Organic solvents are extensively utilized in a range of commercial sectors. Reclaiming and reusing the solvents is also the foremost economically and environmentally useful possibility for managing spent solvents. Purifying the solvents to meet reuse specifications can be challenging. For hydrophilic solvents, water must be removed prior to reuse, yet many hydrophilic solvents form hard-to-separate azeotropic mixtures with water. Such mixtures make separation processes energy-intensive and cause economic challenges. The membrane technologies PV and vapor permeation (VP) can be less energy-intensive than distillation-based processes and have proven to be very effective in removing water from azeotropic mixtures. In PV/VP, separation relies on the solution diffusion interaction between the dense layer of the membrane and therefore the solvent/water mixture. A variability of membrane materials, such as polymeric, inorganic, mixed matrix, and hybrid, has been used for industrial applications. A small set of those commercially obtainable and highlighted here: poly (vinyl alcohol), polyimides, amorphous perfluoro polymers, NaA zeolites, chabazite zeolites, T-type zeolites, and hybrid silicas. Solvents targeted for recovery and then common solvents are chosen for analysis: acetonitrile, 1-butanol, *N,N'*-dimethyl formamide, ethanol, methanol, methyl isobutyl ketone, methyl tert-butyl ether, tetrahydrofuran, acetone, and 2-propanol.

8.8.2   *DAIRY PROCESSING FOR MILK COMPONENTS*

In the last thirty years, membrane processes became major tools within the food trade. This trade represents the second sector of membrane applications, once water treatment, and on equal terms with pharmaceutical and biotechnology applications. Among the food sector, the farm trade has beyond any doubt developed the foremost advanced filtration procedures for concentration and fractionation of molecules from milk and it is derivate. The primary membrane development within the separation procedures of milk elements occurred within the late 1960's with the appearance of membrane separation. The membrane equipment's are enforced throughout the farm process chains: milk reception, cheese-making, whey protein concentration, fractionation of protein, and effluents treatment (Figure 8.1).

Four different membrane operations are distinguished. Ultrafiltration (UF) is the greatest extensively used process in the dairy industry. NF has lately remained established for whey demineralization. Other membrane processes are RO and MF which develop to retain micro-organisms or casein micelles.

1. **Microfiltration (MF):** The MF of milk concerns two main industrial applications:

    i.   **Bacteria Removal from Skimmed Milk:** Similar to most foods, milk provides a good media for spoilage micro-organisms, which might cause alteration of chemical composition of milk and difficulties in milk preservation and transformation. Therein medium frequency projected a noteworthy different to heat-treatment. It is significantly custom-made to the bacterium removal from skim milk; the dimensions of the micro-organisms being within the same vary as fat globules.

    For this application, ceramic membranes used with a pore size of almost 1.4 μm. By combining high cross flow velocity with a low transmembrane pressure, the membranes lead to low matter losses at high fluxes. In large equipment, a second MF stage can be added to concentrate further the first retentate. Due to high retention of bacteria and spores, such a process is used, with some version, by the cheese-makers to produce safe raw milk cheeses.

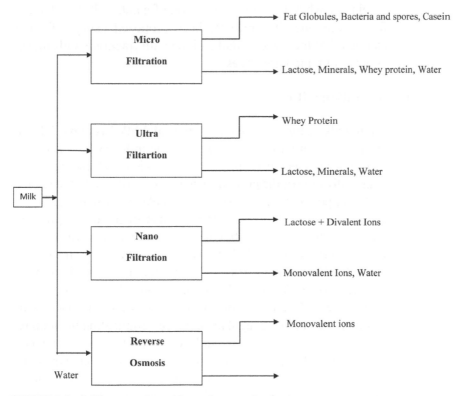

FIGURE 8.1   Milk processing with membrane technologies.

ii. **Selective Concentration:** The permeate, with a composition on the brink of that of a whey, contains whey proteins in their native state, is healthful, free from phase particles and fats, and becomes a helpful fluid to organize whey protein concentrates(WPC).

The retentate is employed for the casein enrichment of cheese milk to boost the cheese-making method, shorter coagulation time and magnified final curd firmness. Additionally, partial removal of whey proteins considerably reduces the prejudicial effects of heat treatment on retentate milk coagulation, and is employed to develop a brand new high-heat milk powder, with a cheese-making ability just like that of raw milk.

Finally, purified casein molecules, obtained by filtration against water, and WPC acquired after UF of milk MF, are exceptional starting fluids for further fractionation of individual caseins or whey proteins. Separation of small milk fat globules from large ones has recently been proposed using MF. It is claimed that the use of small globule with fraction yields makes yogurts soft and hard cheeses.

2. Ultrafiltration (UF):

i. **Standardization and Concentration of Milk Proteins:** UF has been used on milk process plants, for the milk standardization in total macromolecule or protein for drinking and cheese milks. Standardization of the protein content of the milk leads to a regular milk composition all year spherical, freelance of seasonal differences, and so leads to additional constant method and products quality, higher use of existing instrumentality and accrued cheese yields. Spiral wound or hollow ceramic membranes with a molecular cut-off of less than 10 kg. mol^{-1} area unit used for this purpose, at a pressure of 200–400 kPa and flux of 30–120 $L.h^{-1}.m^{-2}$.

UF method permits removal of excess water and lactose by an initial step before coagulation, thereby eliminating the requirement to separate whey from curd. All the proteins are so focused along within the pre-cheese fluid. The acceptable enzymes and micro-organisms are then added into the protein concentrate to coagulate. The benefits of the method for cheese production are nice. The cheese yield is concerning 10–30% above within the ancient method since the whey proteins are enclosed within the pre-cheese. Accelerator usage is usually reduced. The UF method eliminates the requirement for the massive storage tanks historically used for heating and preparation the curds, leading to a saving in each capital investment and energy prices. Visible of the actual fact that the many new cheeses are made exploitation UF.

ii. **Concentration of Whey Proteins:** By far the foremost effective application of membrane processes within the dairy business is that the production of a WPC by UF. UF membranes, with associate acceptable cut-off (10–20 kg mol^{-1}) is familiar take away each the lactose and ions, yielding a retentate with a high protein concentration. The ensuing product will then be more processed by evaporation and spray drying to provide whey proteins caseins (WPC) with various interesting functions. And its purity is ranging from 35 to 85%.

In industrial processes, whey UF is performed in the main in multi-stage spiral wound systems with polysulfone membranes. Processes are presently operated at a temperature of virtually 50°C, requiring a pre-treatment so as to avoid severe fouling throughout operation. Flux is regarding doubly as high as flux at 10°C, which could be a major incentive for operation at such high temperature. However, because of the fast decrease in membrane costs, that alter new systems to work at a temperature of below 15°C, and in respect of microbiological quality of the top product, less temperature method is currently favored.

3.   **Nanofiltration (NF) and Reverse Osmosis (RO):** Concentration of whey on its manufacture site is that the major application of RO attributable to its flexibility and energy consumption compared to vacuum evaporation. The utmost concentration which will be obtained by 18–27% dry out matter, restricted by high pressure, high retentate viscousness, calcium phosphate precipitation and lactose crystallization.

Because of the high salt content of whey, that generates various process difficulties and nutritionary imbalance in food, it became advantageous to remove whey before evaporation. Whey was demineralized within the vary 50–95% by electrodialysis and/or ion exchange, however these operations crystal rectifier to an outsized volume of polluting effluents, high investment and running prices. NF became thus industrial alternative, creating it doable to succeed in at the same time the concentration of dry matter and demineralization (25–50% to even 90% with diafiltration). The method is competitive to RO and electrodialysis. Demineralization is additional selective; NF property remains satisfactory even at the very best demineralization rate; losses of lactose and nitrogen are low. Additionally, the NF step considerably improves the technological characteristics of the concentrate and offers it higher worth.

So membrane operations give the dairy industry with reliable safe, clean, and sober processes. These operations have typically been to substitute, either whole or part, different normal dairy processes and so improve ancient processes. However, additional typically, they need created it doable for brand spanking new and original product to be created that were merely impossible before. Due to the number of analysis that has been targeted on the characterization of milk components, analytical methodologies and separation processes, it is seemingly that the final quality and production potency of the varied milk protein ingredients can increase within the close to future.

8.9 MEMBRANE TECHNOLOGY FOR EFFLUENT TREATMENT

Water contamination by serious metals, cyanides, and dyes is increasing globally and desires to be self-addressed as this may cause water inadequacy in addition as water quality. Totally different techniques are accustomed clean and renew water for human consumption and agricultural functions however they every have limitations. Among those techniques, membrane technology is promising to unravel the problems. Membrane processes have an excellent potential in waste matter treatment to boost treatment potency of waste matter treatment plants. This chapter reviews recent development in membrane technology for waste matter treatment. Differing types of membrane technologies, their properties, mechanisms benefits, limitations, and promising solutions are mentioned.

Several techniques are developed for treatment of wastewater; such strategies embrace RO ion exchange gravity and adsorption among others. Adsorption has been wide wont to take away water contaminants because of its low value, offered of various adsorbents and straightforward operation. Completely different adsorbents that are used embrace use of magnetic nanoparticles activated carbon, nanotubes, and polymer nano-composites; these will take away completely different contaminants together with heavy metals that are terribly harmful even at low concentrations. Although adsorption will take away most of water pollutants, it is some limitations like lack of applicable adsorbents with high adsorption capability and low use of those adsorbents commercially. Thus, there has been a necessity for additional economical techniques like membrane technology. Membrane separation or treatment method chiefly depends on three basic principles, specifically adsorption, and electrostatic phenomenon. The adsorption mechanism within the membrane separation method relies on the hydrophobic interactions of the membrane and therefore the solute. These interactions unremarkably result in additional rejection as a result of it causes a decrease within the pore size of the membrane. The separation of materials through the membrane depends on pore and molecule size. For this reason, varied membrane processes with completely different separation mechanisms are developed. These embrace MF, UF, NF, forward diffusion (FO), and RO.

Membrane processes like MF, NF, UF, and RO are presently used for water use, saltwater, and brackish water. Polymer-based most primarily based membranes are mostly used membrane material however as a result of polymers like polysulfone and polyethersulfone, are hydrophobic, polymeric membranes are cared-for fouling. These results in blockage of membrane pores and decrease membrane performance conjointly will increase operation price by the stringent additional cleanup method.

For polymeric membranes, surface modification of the polymer is important; such surface modification includes grafting, blending, and incorporation of nanomaterials such as TiO_2, ZnO, Al_2O_3, carbon nanotubes (CNTs), and graphene oxide.

From all membrane processes, MF removes very little or no organic matter; but, once pre-treatment is applied, accrued removal of organic material will occur. MF is used as a pre-treatment to RO or NF to reduce fouling potential. The most disadvantages of MF are that it cannot eliminate contaminants (dissolved solids). Also, the NF method is capable of removing ions that contribute extensively to the osmotic pressure hence permits operation pressures that are not up to RO. For NF to be effective pre-treatment is required for a few heavily contaminated waters; Membranes are sensitive to free chlorine. NF membrane for textile sewer water treatment, the ready membrane displayed smart removal of heavy metal ions, common salts, and dyes, showing high removal potency toward metal ions and cationic dyes.

RO is pressure-driven technique wont to take away dissolved solids and smaller particles; RO is simply permeable to water molecules. The applied pressure on RO should be enough so water may be able to overcome the osmotic pressure. The pore structure of RO membranes is way tighter than UF, they convert hard to soft water, and that they are much capable of removing all particles, bacteria, and organics, it needs less maintenance. Some disadvantages embrace the utilization of high pressure, RO membranes are valuable compared to alternative membrane processes and also are tend to fouling. RO has extraordinarily little pores and able to take away particles smaller than 0.1 nm.

Let us see the applicability of membranes separation with selected examples. Two examples are selected for the discussion is ammonia purge stream and acid recovery system:

1. **Ammonia Purge Stream-Prism Membranes:** The membrane processes are the feasible of separation of hydrogen from hydrogenation tail gas, enrichment of refinery gas and air separation. Currently, the growing request of hydrogen in many industrial processes moves forward to hydrogen source and production. However, separation of hydrogen is the most important issue in its production cycle. The first commercial application of membranes in gas separation was the hydrogen separation in the ammonia purge stream by using prism membranes. The gas stream is fed into a membrane gas separation unit with hollow fiber operating at 110–130 bars. The gas components include high concentration of hydrogen (66.5%) and nitrogen (22.2%). Membrane system is able to recover the stream at lower pressure (25–70 bars) with a hydrogen concentration up to 94% and hydrogen recovery up to 90%.

   The prism technology is also applied for other separations, i.e., methanol separation. The methanol/hydrogen stream is further treated in gas separation unit for downstream processing to enrich methanol production in retentate with hydrogen separated in permeate that can be recycled to reactor.

2. **Acid Recovery System:** It is one of the most important systems in effluent technology. Figure 8.2 shows acid recovery system. It's a continuous system of acid recovery in pickling industry. Feed containing acid and salts in water. Almost 7 to 10% HCl acid and 7% salts in feed for recovery system. Process steam (feed) send for pretreatment. Which is primary filtration system where from steam remove the suspended solids. 0.1–0.6% $FeCl_2$ or oil sludge is sent to the roasting unit for formation of oxides from salt at temperature 845°C. This unit produces HCl gas for resorption/reuse and form mineral oxides for reuse. But this roasting unit is optional process in acid recovery system.

After pretreatment sludge free process steam is fed to the first separation unit where dissolved solid contains 10% HCl and 7% salts. In stage, one separation the process is optimized to get desired salt concentration. It can be done by stage-wise to avoid clogging and choking while getting desired salt concentration. Optimized to get desired salt concentration that is 60% is advisable salt concentration is sent back to the roasting unit for avoiding salt crystallization and help for some economical benefits to process.

Then the next salt free process steam is fed to the second separation unit. This unit working with such membrane separation techniques which help to get maximum acid concentration, i.e., up to 20%. It is also done by stage-wise by avoiding clogging and chocking in the process while getting desired acid concentration. This separated acid with concentration of 20% can be recycled to system to getting 25% acid recovery. And at the conclusion of the process, the water can be removed from the entire separation unit with may containing less than 1% acid for precleaning and rinsing before pickling. For better cleaning acidic water is one of the great options.

Example: Continuous system: acid recovery in the pickling industry

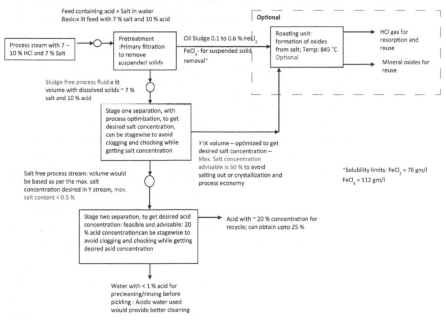

FIGURE 8.2   Acid recovery in the pickling industry.

8.10   ECONOMIC BENEFITS BY COMBINATION WITH MEMBRANE PROCESSES

As discussed in this chapter, membrane technology is very helpful to environment. It helps to reduce the waste, very less consumption of energy though it reduces operating cost, also its environmental friendly and clean technology. Membrane processes show improvement in reaction rate to the formation of desired product. Membrane processes done separations continuously and separation at milder conditions. Easy to recycle and recover the unconverted components with higher conversion and it also helps to improve the product quality. However, the membrane system helps to reduction in waste, losses, and improved component usage for further processes. Automatically these all help to reduction in economy. Some economical benefits with combination of membrane processes are explained with some examples below:

1. **Fermentation-Based Biomolecule Synthesis-Acid, Alcohol, and Biofuel:** Fermentation technology is widely used to produce various economically important compounds which have applications in the energy production, pharmaceutical, chemical, and food industry. Fermentation is highly efficient technique for synthesis of chemicals. It utilizes biological feedstock, provides highly selective functionality. But the issue with the fermentation technique is limitations for higher concentrations and its separation. Also, large waste generated during separations, e.g., separation of LA generates 1 ton of waste / ton LA, and for CA it is 2 ton.

   Lactic acid (LA) is an alpha-hydroxy acid with twin practical teams, creating it appropriate to be used during a sort of chemical transformations and products. LA is employed globally for applications in food, pharmaceutical, textiles, cosmetics, and chemical industries. There are two processes utilized in LA production: chemical synthesis and carbohydrate fermentation. Chemical synthesis produces solely a ceramic mixture of LA using acetaldehyde as a beginning material. Fermentation is most well-liked to supply LA, and concerning 90% of LA is created by fermentation. In fermentation, the substrates for LA production will be renewable and might utilize low-value raw materials like starchy materials, sugarcane, whey, and glycerin. Concerning 90% LA is created by fermentation technique however, this method has not been additionally profitable as a result of the high value in separation method. To deal with this downside, the event of additional economical and viable separation technologies is required. Therefore, that is completed with membrane separation.

   Since 1960, membrane separation processes have been recommended as a substitute for LA extraction. Advances in separation and purification based on membranes, particularly MF, UF, and electrodialysis technology allowed the manufacture of new production processes, which do not produce a salt residue, instead of the traditional precipitation process of LA.

Electrodialysis is a separation method employing selectively permeable cationic and anionic exchange membranes. These membranes are alternately arranged between the cathode and the anode; by spread over an electric potential between the electrodes, cations migrate to cathode and anions to the anode. Membranes used in electrodialysis are polymeric, non-porous, and have a thickness between 10 and 500 μm. Electrodialysis is applied to remove salts from solutions or to concentrate ionic substances. Membranes can operate at approximately 80% of theoretical thermodynamic efficiency.

So this concludes that the precipitation is that the typical technique for LA recovery, however it is unattractive for economic and environmental viewpoint as a result of it generates LA with low purity and solid waste residue. Also, solvent extraction has the advantage of not generating heaps of waste because of the precipitation technique. However, it needs high exchange space, high price with solvent recovery, and high toxicity of extractants, that limits the large-scale potential. Advances in solvent extraction, notably with the novel extractants, have given promising results. Oppositely, membrane-based separation technologies have according to high property, high levels of purification and separation. Despite its benefits, high price of membranes, polarization, and fouling issues forestall the utilization of those processes.

2. **Ethanol Concentration-Absolute Ethanol Formation:** Most of the world's ethanol is produced by fermentation of crops (93%) with artificial ethanol (7%) being produced by direct hydration of ethane. The fermentation of plant material (e.g., barley and rice) is the route by which alcoholic drinks (e.g., beer, whiskey, gin, and vodka) are produced. It is also how bioethanol for biofuels is produced. Of the uses of bioethanol, easily the greatest significant fuel for cars. However, an increasing one is in the manufacture of ethane as a route to poly (ethane). The other main uses of ethanol as a chemical intermediate are for:

• glycol ethers;
• ethanolamine/ethylamine; and
• ethyl propenoate.

It is conjointly used as a solvent within the manufacture of cosmetics, pharmaceuticals, detergents, inks, and coatings. While bioethanol is taken into account as a sustainable energy supply, it needs any purification for uses aside from fuel. The foremost common purification technique used within the fermentation alcohol business is rectification by any distillation. However, distillation has vital disadvantages as well as high value and restricted separation capability. Although associate azeotrope could be a technique for separation with constant-boiling mixture within which the composition of the vapor is that the same as that of the liquid. Thus, the two parts cannot be separated by fractional distillation process. As imagine, it is very tough to distil this kind of substance. In fact, the foremost targeted sort of fermentation alcohol, associate azeotrope, is around 95.6% ethanol by weight as a result of pure ethanol is essentially nonexistent. Azeotropes exist in resolution at a boiling purpose specific for that part.

The separation of ethanol-water mixtures is of great significance for the production of ethanol from biomass. Together UF and PV processes are used for the continuous processing of fermentation and separation, the removal of ethanol from the UF permeate can be accomplished by PV. Separation of ethanol-water mixtures by the PV process is examined. The preparation of several membrane types (homogeneous, asymmetric, and composite) are used in PV separation technology. Dissimilar types of membrane provide altered characteristics. The self-supported polymeric membranes may provide an economic attraction in membrane production, but not in performance, due to their poor mechanical properties, low chemical resistance, and swelling. The supported membranes improve mechanical strength and separation performance. The inorganic membranes provide an outstanding separation performance, but are extremely difficult in process ability, leading to a higher price in industrial usage. The mixed matrix membranes thus provide an attractive alternative that are able to overcome all barriers while achieving both permeation flux and separation factor.

8.11   WAY INTO FUTURE-SCOPE AND BENEFITS

The discussion up till now shows large growth potential and applicability of membrane technology in green technology applications. Though multiple applications are available in current date, their applicability needs to be established with the help of research and developments and its implementation in real life situations. For that purpose, some of the targets have to be established and reached as described below:

1. **Research Targets:** Technological advances towards process and energy economy: The separation parameters and separation efficiency need to be optimized to improve the material quality. Such improved quality material can be utilized in further processes and can provide techno-economical benefits. Additionally, the research needs to be progressed for improvement in energy economy. As per an estimate, separation processes are major energy-intensive systems in process design and engineering. A reduction in energy consumption for separation can be highly beneficial towards the goal of energy economy in green technology.

    Further process modifications can be targeted towards improvement in solvent systems, which would provide efficient heat and mass transfer. This would be beneficial in improving energy economy along with processing benefits. An improvement in mass transfer can affect the reaction kinetics and process conversions. Major issues with the mass transfer controlled reactions are catalytic reactions. An improved catalyst with feasibility of its easy intrinsic separation would provide large benefits. Many of the reactions are either equilibrium reactions and require conversions after product formation. The unconverted reactants are recycled back to the system. This causes loss of energy, reaction kinetics, and conversion. Additionally use of chemicals for separation and recovery unreacted component might result in waste generation. This needs *in situ* separation of desired components. It would provide solution towards many of these issues. A carefully designed membrane system can be combined with conventional processing for *in situ* separation; can be useful in drawing these technoeconomical benefits. These types of *in situ* separations are highly essential for materials with sensitive functional groups. Their separation and further utilization would avoid need of chemical processing and contamination. Thus, avoid degradation of functionality and loss of product properties.

    In the designed membrane system, development of some synthetic methodologies is also help to technoeconomical assistances. These synthetic methodology supports to make atom efficient and benign to human health/environment. According to membrane process designed for intensive manufacturing of chemicals which required less energy to process for separation of molecules. And some of these molecules are sent back to the process. This recyclability of molecules increases the product quality and reduction in large economical requirements.

Today the waste streams of energy-producing units are some of the major polluting factors. The wastes from thermal or nuclear power plants are harmful for humans and environment based upon their sheer volume and properties. Membrane-based systems like fuel cells can be an option, which are reported to produce negligible amount of waste. Though the processes are well reported a large investigation are needed for their scale up and commercial feasibility.

2. **Implementation:** Though multiple steps are being observed towards research and development, their practical applicability is still limited. Industry is still reluctant and skeptical about accepting the technological development. The steps needed to be taken by industries and enforcement agencies for implementation of the technological advances. It can be done by taking the following measures:
   a. The steps of the process can be redesigned for improved adoption, quality enhancement, and product improvements.
   b. Develop the research programs to facilitate technology transfer.
   c. Designed and manipulated the systematic recognition of hazard/toxicity as property of molecular structure.
   d. Development and utilization of practical investigations for design and manipulation with green technology principles.

Implementation of some of these steps would help us to make the systems better.

KEYWORDS

- **economical separation**
- **effluent treatment**
- **green technology**
- **membrane separation**
- **process intensification**
- **recovery and recycle**

REFERENCES

1. Angelo, B., & Fausto, G., (2011). *Membranes for Membrane Reactors: Preparation, Optimization, and Selection*. John Wiley & Sons, UK.
2. Baker, R. W., Cussler, E. L., Eykamp, W., Koros, W. J., Riley, R. L., & Strathman, H., (1991). *Membrane Separation Systems: Recent Developments and Future Directions*. Noyes Data Corporation, USA.
3. Fendler, J. H., (1994). *Membrane-Mimetic Approach to Advanced Materials*. Springer-Verlag Berlin, Heidelberg.
4. Grandison, A. S., & Lewis, M. J., (1996). *Separation Processes in Food and Biotechnology Industries: Principles and Applications*. Wood head Publishing Ltd. England.
5. Ian, D. W., Edward, R. A., Michael, C., & Colin, F. P., (2000). *Encyclopedia of Separation Science*. Elsevier Science Ltd., Academic Press, UK.
6. Marcel, M., (1996). *Basic Principals of Membrane Technology*. Kluwer Academic Publishers, Netherlands.
7. Mark, C. P., (1990). *Handbook of Industrial Membrane Technology*. Noyes Publications, USA.
8. Mel, S., (2002). *Encyclopedia of Smart Materials*. A Wiley-Interscience Publication, John Wiley & Sons, Inc. USA.
9. Michael, C. F., & Stephen, W. D., (1999). *Encyclopedia of Bioprocess Technology: Fermentation, Biocatalysis, and Bioseparation*. A Wiley-Interscience Publication, John Wiley & Sons, Inc. Canada.
10. Neogi, P., (1996). *Diffusion in Polymers*. Marcel Dekker Inc. USA.
11. Nicholas, P. C., (1996). *Biotechnology for Waste and Wastewater Treatment*. Noyes Publications, USA.
12. Nunes, S. P., & Peinemann, K. V., (2001). *Membrane Technology in Chemical Industry*. Wiley-VCH Verlag GmbH.
13. Paul, T. A., & Mary, M. K., (2002). Origins, current status, and future challenges of green chemistry. *Acc. Chem. Res., 35*, 686–694.
14. Richard, W. B., (2002). *Membrane Technology and Applications*. John Wiley & Sons, Ltd., USA.
15. Winston, H., & Sirkar, K. K., (1992). *Membrane Handbook*. Van Nostrand Reinhold, Springer USA.

CHAPTER 9

Green Hydrogen Energy: Storage in Carbon Nanomaterials and Polymers

BRAHMANANDA CHAKRABORTY[1] and GOPAL SANYAL[2]

[1]*High Pressure and Synchrotron Radiation Physics Division, Bhabha Atomic Research Center, Trombay, Mumbai – 400085, India*

[2]*Mechanical Metallurgy Division, Bhabha Atomic Research Center, Trombay, Mumbai – 400085, India*

ABSTRACT

One of the greatest challenges of this century is to find an alternative fuel for vehicular propulsion which can replace widely used fossil fuels. Hydrogen is one of the finest choices due to its clean energy and highly plentiful in nature. And also, it has the highest energy value per unit weight ($142 \, MJ \, kg^{-1}$) compared to liquid hydrocarbons ($47 \, MJ \, kg^{-1}$). The conventional hydrogen storage technologies, namely, high-pressure tank and liquid state storage, are not applicable due to large size and higher energy cost for liquefaction. Solid-state storage may become a viable technology provided the storage medium can absorb a large amount ($\sim$6.5 wt%) of hydrogen and can release them easily as recommended by the department of energy (DoE), USA. Recently, carbon nanomaterials have drawn immense attention from researchers due to their high surface area, high electrical and thermal conductivity, and high mechanical strength. Although for pure carbon nanostructures, the physisorption of hydrogen is negligible at room temperature, the functionalized carbon nanomaterials are the promising candidates for hydrogen storage media at ambient conditions. This chapter describes the issues and challenges for hydrogen storage in functionalized carbon nanomaterials. It will highlight the bonding, charge transfer mechanism, and hydrogen storage capability of novel carbon nanostructure (carbon nanotube, graphene, and graphyne).

Polymers are also potential hydrogen storage materials due to their pure organic nature, tunable structures, and large scale synthesis. This chapter discusses about different types of polymers such as conjugated microporous polymers, porous polymers networks, hyper cross-linked polymers, polymers of intrinsic microporosity (PIM) and their hydrogen storage capabilities.

9.1 INTRODUCTION

Everything is energy and that's all there is to it. Fuel is invariably the precursor to energy. Conventional fossil fuels are at the verge of culmination through rapid depletion with progress of civilization [1, 2] with a daily global consumption of over 89 million barrels they suffer the largest drawback being rapidly exhaustive, causing huge pollutant emission and volatile in price-controlled by political factors [3]. Essentially, that thrusts people exploring alternative fuels such as bio-diesel, bio-alcohol, hydrogen, etc. for sustaining advancement [4–6]. Precisely, rise in price and negative impact on environment for depleting fossil fuels are the chief motivation behind development of alternative fuel technologies (Figure 9.1(a)). On the other hand, alternative fuels are obviously aimed at reduction in fuel cost and emission, and off course a cleaner environment. Among all alternative fuels, hydrogen is a promising energy carrier with zero-emission while burnt with oxygen. Unlike conventional liquid fuel, it can be generated from a broader spectrum of primary stationary energy resources including nuclear and solar power being cleaner than fossil fuel; and in turn, the produced hydrogen can be utilized for generating electricity (Figure 9.1(b)) [7], or fueling vehicles either through direct burning inside an internal combustion engine (ICE) or as precursor for fuel cells [3, 8, 9]. The energy released through oxidation enables hydrogen to act as a fuel. Hydrogen storage is indispensable machinery for the progress of hydrogen and fuel cell skills in applications encompassing stationery as well as portable power [10–14]. Storage of hydrogen is one of the major technical obstacles which are being addressed by researchers worldwide for the last a few decades [15–17]. The following section would discuss about importance of hydrogen energy in brief. Thereafter, issues, and challenges regarding hydrogen storage would be narrated in detail including an ephemeral insight about various hydrogen storage means. The succeeding three sections would focus on progress in research about hydrogen storage within carbon nanomaterials and organic polymers highlighting theoretical insights behind the fundamental concepts responsible for the phenomena. Finally, the chapter would be concluded elucidating future research in this direction.

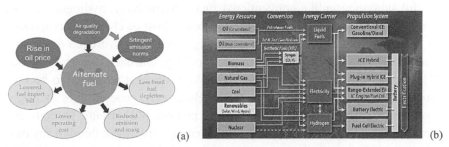

(a)   (b)

FIGURE 9.1 (a) Alternate fuels' emergence and benefits, and (b) transformation of stationary power into transportation system.

Source: b) Reprinted with permission from National Research Council [7]. © 2008 National Academies Press

9.2 HYDROGEN FUEL: IMPORTANCE, MERITS, AND DEMERITS

Hydrogen is an energy carrier which can be utilized for a broad range of applications [18–20]. Hydrogen promises huge long-term potential for utilization as fuel in usage, where presently fossil fuels are in practice-precisely for driving the vehicles, which would tender immediate benefits in terms of curbed pollution and fresher environment (Figure 9.2(a)) [21]. Among all fuels, hydrogen has the highest gravimetric energy density of 143 MJ kg^{-1} and octane number more than 130 [3]. The rock bottom flashpoint of –231°C designates hydrogen easily burnable on one hand and highly fire hazardous on the other. However, auto-ignition temperature of hydrogen is still quite high (585°C). Besides usage in automobile fuel, it has commenced to be utilized in commercial fuel cell vehicles and spacecraft propulsion [22, 23]. A fuel cell coupled with an electric motor is much more efficient than a conventional ICE operating on gasoline [24–26]. Again, hydrogen itself can serve as fuel for ICEs. Energy generated from 1 kg of hydrogen is equivalent to that generated from 2.8 kg of gasoline [27]. The world's dependence on hydrogen energy exhibits a steady rising trend since the past a few decades (Figure 9.2(b)) [28].

Molecular hydrogen is unfortunately unavailable on earth in expedient innate reservoirs [29]. Most hydrogen on globe is bonded to oxygen in water and to carbon in surviving or dead and/or fossilized biomass. Hydrogen must thus be separated [19] from any of its compound using primary energy source such as solar energy [30], biomass [31], electricity [32], or hydrocarbons [33] (Figure 9.1(b)). If it is produced by splitting water into hydrogen and oxygen [34–36], water is again obtained, when hydrogen is burnt without any formation of any harmful substance unlike conventional fuel. Alternatively, hydrogen may be obtained by breaking the chemical bonds from a host of chemical compounds [39, 40]. Conventional natural gas reforming process for hydrogen production induces significant environmental impacts for huge emission of carbon dioxide [41]. Greener processes are still not technologically that advanced to become commercially viable at large scale. Comprehensively, generation of hydrogen is yet an expensive and energy-intensive process [42]. Being gaseous and extremely lightweight, it is not easy to store enough hydrogen through any easy and convenient way for usage as per needs. In other words, containing good amount of hydrogen is a costly affair till date for its escaping tendency. Hence, researchers have been endeavoring to find suitable storage means for hydrogen especially for mobile applications since past a few decades.

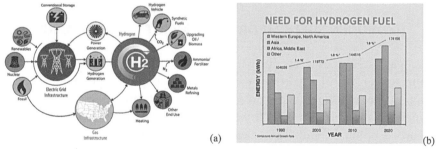

FIGURE 9.2 (a) Potential for wide-scale hydrogen production and utilization (From https://www.energy.gov/eere/fuelcells/h2scale, with permission), and (b) worldwide emerging demand of hydrogen fuel (From https://www.slideshare.net/SridharSibi/hydrogen-fuel-amp-its-sustainable-development, with permission).

9.3   HYDROGEN STORAGE: CHALLENGES AND ISSUES

Being carrier of energy, hydrogen storage technically means storing energy. Depending on functionality, hydrogen storage technologies may be classified into five major groups: (1) high-pressure gas storage, (2) liquid hydrogen, (3) physically bound hydrogen, (4) chemically bound hydrogen, and (5) hydrolytic evolution of hydrogen [11]. In view of application, all hydrogen storage technologies fall into either of these two domains: stationary storage and portable storage or on-board storage or storage for mobile applications. High pressure gas storage is regarded as the most common hydrogen storage solution till date. Liquid hydrogen although being quite mature technology, needs very low temperature for liquefaction. Among all forms of hydrogen, liquid hydrogen bestows both the highest volumetric and gravimetric storage density (Figure 9.3(a [43], b [44])) The above two solutions are very much suitable for stationary applications such as underground gas storage to provide grid energy availability for intermittent energy sources or 'power to gas' technology to translate electrical power to a gas fuel. Hydrogen energy density is not a great concern for stationary applications.

On the other hand, for on-board hydrogen storage for fueling a vehicle, energy density is very much critical. The existing options for hydrogen storage need large storage volume that turns them impractical for portable as well as stationary applications. Portability is one of the greatest challenges in the automotive engineering, where high density storage systems are infected with problems for safety concerns. Fuel cell-powered motor vehicles are required to offer driving range, more than 300 miles and this is impossible to attain with traditional storage methods. Secondly, both liquefaction and gas pressurization are too expensive being very much energy-intensive for mobile applications. Physically bound hydrogen essentially refers to adsorption of molecular hydrogen at solid surface through Van der Waals interaction. Hydrogen in that case is thought to be stored within that matter. Similarly, chemically bound hydrogen is that which is captured through chemical reaction with the carrier material. Hydrolytic evolution of hydrogen refers to any chemical or electrochemical process involving water, capable of releasing hydrogen. These technologies are collectively termed as storing hydrogen in solid state. Most current research into hydrogen storage for mobile applications is concentrated on accumulating hydrogen as a low-weight, compact energy carrier. A long-term objective set by the Fuel Cell Technology Office engrosses usage of nano-materials to perk up maximum range. The DoE has set the aims for onboard hydrogen storage for light cars (Table 9.1) [45]. The list of necessities includes parameters involving operability, volumetric, and gravimetric capacity, cost, and durability. These targets have been placed as the goal for a multiyear research plan anticipated to offer a substitute to fossil fuels. These targets or required properties are closely linked with three sets of parameters: structure (crystalline form, particle size or specific surface area, and structural defects), chemistry (components, phases, composition, catalysts, and impurities), and reaction or diffusion path (elementary reactions, transient species, etc.). These are mutually interacting and the critical issues vary for different types of hydrogen storage materials (Figure 9.3(c)) [46].

Solid-state hydrogen storage essentially requires hydrogen gas to be in contact with solid. The pressure can be markedly lower compared to gas storage but will evidently depend on the operative features of the storage technology. Particularly, service temperature ranges distinguish solid storage from the gaseous one. Control of adsorption-desorption cycles in solid storage devices is most often managed thermally, hence operating temperatures could surpass the near-ambient temperatures emblematic for hydrogen gas storage. The volumetric and gravimetric storage densities are directly influenced by hydrogen storage material properties together with physical, thermodynamic, and chemical/kinetic. Extra influences on energy densities include, but not limited to, properties of the containment material, operational environment, available energy, and required hydrogen delivery rates (Figure 9.3(d) [11]). Two more concerns are often associated with solid-state hydrogen storage devices. The first one is stress in the containment material persuaded by expansion of the solid storage media during absorption or desorption of hydrogen. Such stresses may, in principle, be rather large and should be well-thought-out in addition to the stress associated with gas pressure. The second consideration is probable chemical interaction between the solid storage media and containment material that could lead to poisoning, corrosion of the solid storage media, and aggravated hydrogen embrittlement of the structural metal.

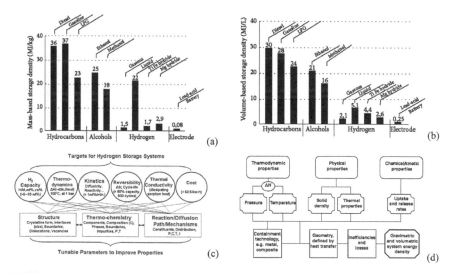

FIGURE 9.3   (a) Comparison of mass-based storage density of hydrogen with other fuels (From Zacharia [43], with permission), (b) comparison of volume-based storage density of hydrogen with other fuels (Reprinted with permission from Das [44]. © 1996 Elsevier), (c) Practical targets for on-board hydrogen storage in relation to key materials parameters (Reprinted with permission from Guo, Shang, and Aguey-Zinsou [46]. © 2008 Elsevier), and (d) the hierarchical influence of various hydrogen storage material properties on the system characteristics and efficiencies (Reprinted with permission from Walker [11]. © 2008 Elsevier.)

TABLE 9.1   Technical System Targets: Onboard Hydrogen Storage for Light-Duty Fuel Cell Vehicles

Storage parameter	Units	2020	2025	Ultimate
System Gravimetric Capacity				
Usable, specific energy from H$_2$ (net useful energy/max system mass)	kWh/kg (kg H$_2$/kg system)	1.5 (0.045)	1.8 (0.055)	2.2 (0.065)
System Volumetric Capacity				
Usable energy density from H$_2$ (net useful energy/max system volume)	kWh/L (kg H$_2$/L system)	1.0 (0.030)	1.3 (0.040)	1.7 (0.050)
Storage System Cost				
Storage system cost	$/kWh net ($/kg H$_2$)	10 (333)	9 (300)	8 (266)
Fuel cost	$/gge at pump	4	4	4
Durability/Operability				
Operating ambient temperature	°C	−40/60 (sun)	−40/60 (sun)	−40/60 (sun)
Min/max delivery temperature	°C	−40/85	−40/85	−40/85
Operational cycle life (1/4 tank to full)	cycles	1500	1500	1500
Min delivery pressure from storage system	bar (abs)	5	5	5
Max delivery pressure from storage system	bar (abs)	12	12	12
Onboard efficiency	%	90	90	90
'Well' to power plant efficiency	%	60	60	60
Charging/Discharging Rates				
System fill time	min	3–5	3–5	3–5
Minimum full flow rate (e.g., 1.6 g/s target for 80 kW rated fuel cell power)	(g/s)/kW	0.02	0.02	0.02
Average flow rate	(g/s)/kW	0.004	0.004	0.004
Start time to full flow (20°C)	s	5	5	5
Start time to full flow (−20°C)	s	15	15	15
Transient response at operating temperature 10%–90% and 90%–0% (based on full flow rate)	s	0.75	0.75	0.75

TABLE 10.1   *(Continued)*

Fuel Quality				
Fuel quality (H₂ from storage)	%H₂	Meet or exceed SAE J2719		
Dormancy				
Dormancy time target (minimum until first release from initial 95% usable capacity)	Days	7	10	14
Boil-off loss target (max reduction from initial 95% usable capacity after 30 days)	%	10	10	10
Environmental Health and Safety				
Permeation and leakage	-	Meet or exceed SAE J2579 for system safety		
Toxicity	-	Meet or exceed applicable standards		
Safety	-	Conduct and evaluate failure analysis		

Useful constants: 0.2778 kWh/MJ; Lower heating value for H2 is 33.3 kWh/kg H2; 1 kg H2≅ 1 gal gasoline equivalent (gge) on energy basis.

*Source:*https://www.energy.gov/eere/fuelcells/doe-technical-targets-onboard-hydrogen-storage-light-duty-vehicles. With permission.

9.4   CONVENTIONAL AND ADVANCED MATERIALS FOR HYDROGEN STORAGE

There is a host of materials which are in developmental stage for on-board usage (Figure 9.4(a)). For example, chemical storage could tender high storage performance for strong binding of hydrogen and high storage densities. But regeneration of storage material is still a big issue. Similarly, as an alternative to chemical storage mentioned above, there are many physical means with which hydrogen can be stored. Hydrogen remains in that case in physical forms, i.e., as gas, adsorbate, molecular inclusions, or supercritical fluid. Theoretical limitations, as well as experimental findings, are considered [47] concerning the gravimetric and volumetric capacity of the storage media, as well as safety, refilling-time demands, etc. The following discussion shall be restricted to only important solid-state gas storage materials having promising track record or potential for on-board usage.

A huge number of chemical storage systems are under study, involving hydrogenation/dehydrogenation reactions, hydrolysis reactions, ammonia,

ammonia borane and other boron hydrides, alane or aluminum hydride, etc., [48]. The most promising chemical approach is invariably the electro-chemical hydrogen storage, since release of hydrogen may be controlled by applied electricity. For example, metal hydrides, such as LiH, MgH_2, $NaAlH_4$, $LiAlH_4$, $LaNi_5H_6$, $TiFeH_2$ and palladium hydride, with changeable degrees of efficiency, can be utilized as a storage method for hydrogen, often reversibly [49]. These materials have great energy density, although their specific energy is habitually worse than the foremost hydrocarbon fuels (Figure 9.4(b)) [50]. Nano-metal hydrides have a number of properties that turn them even better contenders for future hydrogen storage systems [51]. At nanoscale, chemical, and structural properties such as sorption site density and particle size demonstrate a significant improvement in sorption kinetics and temperature for hydrogen diffusion or release. However, the downsides of nanoscale materials comprise poor total sorption capacity and heat transfer. Hydrazine is a promising non-metal hydride for ease in handling and storage. Aluminum has been projected as an energy storage method by many researchers [52, 53]. Hydrogen can be obtained from aluminum by reacting it with water. Before that, aluminum must be exposed from its natural oxide layer. That process requires pulverization and chemical reactions with caustic substances. The byproduct, aluminum oxide, can be recycled back into aluminum with the well-known Hall-Héroult process, causing the process theoretically renewable. Graphene is chemically effi-cient to store hydrogen efficiently. Upon hydrogenation, graphene becomes graphane. It releases stored hydrogen at 450°C [54].

Physisorption of gases is happened by the interplay between surface area and variations in entropy and enthalpy. Though surface area is frequently used as a proxy for adsorption sites, determination of exactly available surface area for porous materials is difficult. The enthalpy of adsorption determines the temperatures at which reversible adsorption happens. For physisorption at supercritical temperatures, the density of stored hydrogen is predictable to be lower than that of liquid hydrogen. Although this limitation is not valid when chemisorption or a chemical reaction happens between hydrogen and a substrate, it is relevant to most of nanoporous hydrogen storage materials and may be used to define a saturation limit for hydrogen adsorption in such materials. Activated carbons are essentially amorphous carbon mate-rials having high porosity with very high apparent surface area. Hydrogen physisorption can be raised in these substances by optimizing pore diameter to around 7 Å with a corresponding increase in apparent surface area [55]. They can be recycled from waste materials like cigarette butts, indicating economic mileage [56]. A clathrate is a chemical substance having a lattice that contains or traps molecules. They are able to cage hydrogen but need very high pressure for stability. Tetrahydrofuran has been demonstrated to act as a catalyst to reduce the pressure requirement for clathrate hydrate [57]. However, they are yet to achieve DoE's target (Figure 9.4(c)) [58]. Metal-organic frameworks (MOFs) represent another class of synthetic porous matter that store energy in terms of hydrogen and at molecular level [59, 60]. These are highly crystalline organic-inorganic hybrid structures that include organic ligands as linkers and metal clusters or ions (secondary building units or SBU) as nodes. When guest molecules or solvent dwelling in the pores are detached during the process of solvent exchange through heating under vacuum, MOFs' porous structure may be achieved without subverting the frame and hydrogen molecules will be adsorbed onto the pores' surface by physisorption. Glass capillary arrays [61] and glass microsphere [62] are other notable technologies in this direction.

Apart from cost, operability, and durability issues, the three limiting factors for the use of hydrogen fuel cells (HFC) embrace size, efficiency, and safe onboard storage of the gas. To address these issues, the use of nanomaterials has been proposed as an alternative option to the traditional hydrogen storage systems. The use of nanomaterials could endow with a higher density system which is expected to enhance the driving range limit (300 miles) set by the DoE. Carbonaceous materials, e.g., graphite, carbon nanotubes (CNTs), and metal hydrides are the major focus of researchers in this direction. Yu et al. pictorially formulated schemes of three different approaches for nanomaterials to store hydrogen (Figure 9.4(d)) [63]. Carbonaceous materials are at present being considered for on-board storage systems for their multifunctionality, robustness, versatility, suitable mechanical properties and low cost comparing other alternatives [64]. However, low hydrogen capacity of CNTs (~3.0–7.0 wt% at 77 K) and high release temperature of fullerene (600°C) does not attract pure carbon nanomaterial as promising candidate for hydrogen storage. It is pertinent to note here that incorporation of nano-materials within carbonaceous materials through chemical functionalization is proven not only to be beneficial in enhancing hydrogen storage capacity but also are helpful in enrichment of other performances such as transport and catalytic properties (e.g., Nafion membranes) through improvement in hydrogen splitting kinetics by virtue of catalytic activity of nano-particles (e.g., TiO_2/SnO_2) [64]. In essence, another such promising class of materials is the organic polymers containing nanosized pores. They have added weight-advantage being comparatively lightweight. Like carbon nanomaterials, their hydrogen storage performance is also not appreciable at ambiance; however, addition of little quantum nanomaterials causes significant improvement in their storage performance. Hence, decoration of pure carbonaceous nanomaterials and organic polymers with suitable nanoparticles has virtually unlimited potential in enhancing their hydrogen storage performance.

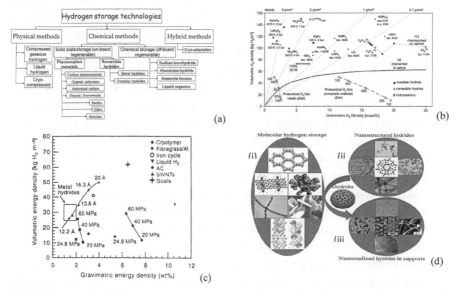

FIGURE 9.4 (a) Different hydrogen storage methods, (b) Comparison of volumetric hydrogen storage density against gravimetric for different pure storage materials (Reprinted with permission from Klell [50]. © 2010 Wiley-VCH Verlag GmbH & Co.), (c) Installed energy densities for several vehicular hydrogen storage technologies (Reprinted with permission from Dillon et al. [58]. © 1997 Springer Nature), (d) Scheme of three approaches for nanostructured materials applied for hydrogen storage. (i) Several typical nanostructures (porous carbon, graphene, BN nanotube, MOF, COF, and CMP) for molecular hydrogen storage; (ii) several nanosized hydrides (MgH2, Mg(BH4)2, Pd, Li2NH); (iii) several hydrides nanoconfined in supports (MgH2@MOF, MgH2@graphene, MgH2@carbon, NaAlH4@ SBA-15, LiNH2@carbon) (Reprinted with permission from Yu et al. [63]. © 2017 Elsevier.)

9.5 HYDROGEN STORAGE IN CARBON NANOMATERIALS

Carbon nanomaterials are useful for effectual energy translation and storage. They possess unique size-/surface-dependent (e.g., morphological, mechanical, optical, and electrical) properties responsible for their remarkable performance for storing hydrogen at high pressure and low temperature in pure form by virtue of intercalation. The carbon-based materials have received outstanding consideration as potential storage materials for their low cost, good recycling characteristics, easy accessibility, low densities, and reasonably good chemical stability, wide diversities of bulk and pore structures, and amenability to synthesize variants or post-synthetically engineer traits using an extensive range of manufacturing, activation, and carbonization routes. A number of theoretical investigations confirm that proper incorporation of selective elemental nanoparticles within carbon lattice causes change in binding energy and desorption temperature resulting in great storage of hydrogen at ambient condition feasible. However, experimental efforts are still not able to tally with all theoretically ambitious predictions for practical limitations such as oxidation of decorated nano-particles, agglomeration, uneven distribution, etc. The science of the hydrogen multilayer intercalation with carbonaceous nanostructures can be understood well and further enriched in the light of so far available analytical and observational research outputs. On the other hand, many researchers do not believe in a leeway of a physical-chemical interpretation and a practice comprehension of the hydrogen multilayer intercalation with carbonaceous nanostructures at technological pressures and ambient temperature. Hence, it seems valuable to deepen and broaden the discussion in scientific publications of this genuine open question, and/or international cooperation. The following subsections essentially review progress in the research and development of carbon nanomaterials during the past two decades for hydrogen storage, along with some discussions on contests and standpoints regarding improvement in hydrogen storage performance through decoration with nanoparticles.

9.5.1   CARBON NANOMATERIALS: A BRIEF OVERVIEW

Carbon, having atomic level ability to polymerize, is one of the most appealing elements, with the capability to form an extensive range of structures, often with fundamentally different properties. Orthodox examples of carbon allotropes encompass "hard" diamond and "soft" graphite used in a wide range of products, including consumer goods in a range of human activity [65]. All nanomaterials composing of carbon atoms are termed as carbon-based or carbon nanomaterials. Classification of carbon-based nanomaterials is frequently performed as per geometrical structure. Manipulation of atomic bonding and shape of a single atomic layer of graphite may lead to formation of all different carbon nanomaterials' allotropes as shown in Figure 9.5(a). Carbon nanostructures comprise particles that can be tube-shaped, horn-shaped, spherical, or ellipsoidal. Nanoparticles having the shape of tubes are termed as CNTs whereas horn-shaped particles are nano-horns or spheres or ellipsoids belong to the group of fullerenes (Figure 9.5(b)) [66].

Fullerenes, discovered in 1985, are an allotropic alteration of carbon, often termed as a molecular form of carbon, or carbon molecules [67]. The fullerene family comprises a number of atomic C_n clusters ($n > 20$), composed of carbon atoms on a spherical surface. The carbon atoms are typically located on the surface of the sphere at the vertices of hexagons and pentagons. The carbon atoms are habitually present in the sp^2-hybrid form and connected together by covalent bonds. Fullerene C_{60} is the most familiar and best-investigated fullerene. The spherical molecule is highly symmetric and contains 60 carbon atoms, situated at the vertices of twelve pentagons and twenty hexagons. The diameter of fullerene C_{60} is 0.7 nm [68].

CNTs, discovered in 1991, are one of the carbon allotropes with outstanding properties suitable for technical applications. These are characterized by cylindrical structures with a diameter of numerous nanometers, consisting of rolled graphene sheets. These may vary in length, diameter, chirality (symmetry of the rolled graphite sheet) and the number of layers. As per their structure, CNTs may be classified into two major groups: single-walled nanotubes (SWCNTs) and multi-walled nanotubes (MWCNTs). Some researchers in addition isolate double-walled carbon nanotubes (DWCNTs) as a detach class of CNTs. Generally, SWCNTs have a diameter around 1 to 3 nm and a span of a few micrometers. MWCNTs have a diameter of 5 to 40 nm and a length around 10 μm. However, recently synthesis of CNTs with a length of yet 550 mm has been reported [69]. The organization of CNTs leads to excellent properties with a sole combination of elasticity, strength, and rigidity compared with other fibrous competitors. For instance, CNTs display considerably higher aspect ratios (length to diameter ratios) than other materials and superior aspect ratios for SWCNTs as compared with MWCNTs due to their smaller diameter. In addition, CNTs show high electrical and thermal conductivity compared to other conductive resources. Electrical properties of SWCNTs depend on their chirality, i.e., hexagon orientation with respect to the tube axis. Consequently, SWCNTs are classified into three sub-classes: (i) armchair (electrical conductivity > copper), (ii) zigzag (semiconductive properties), and (iii) chiral (semiconductive properties). By contrast, MWCNTs having multiple carbon layers, often with variable chirality, can exhibit astonishing mechanical properties instead of exceptional electrical characteristics.

Graphene is a two-dimensional allotropic form of carbon, shaped by single layers of carbon atoms [70, 71]. Here, carbon atoms exhibit sp^2-hybridization linked by σ- and π-bonds in a two-dimensional hexagonal crystal lattice with a separation of 0.142 nm between adjoining atoms of carbon hexagons.

Graphene also essentially represents a structural element of some other carbon allotropes, such as graphite, fullerenes, and CNTs. Theoretical studies on graphene started a long time before the real material samples were obtained. Comprehensive studies on the properties of graphene are still ongoing. Graphene has many exclusive physical properties, such as very high mechanical rigidity and an elevated thermal stability. The electric properties of this carbon allotrope are fundamentally dissimilar to the properties of three-dimensional materials. Though theoretically physisorption (0.01–0.06 eV) of molecular hydrogen on graphene sheet is much less energy-intensive than chemisorption ($\sim$1.5 eV), the achievable gravimetric density through chemisorption (8.3%) is much higher than physisorption (3.3%) (Figure 9.5(c)) [72]. Carbon nanotube can be considered as a curved form of graphene having added advantage of more hydrogen storage potential for its curvature.

Besides these three real carbon nanomaterials mentioned above, there are a number of theoretical two-dimensional carbon nanomaterials such as graphyne [73], graphdiyne [74] pentagraphene [75], etc., having superior hydrogen storage properties under study.

9.5.2   HYDROGEN STORAGE IN PURE CARBON NANOMATERIALS

The physisorption of hydrogen along high-surface carbons has clarified one focal point in view of ease of hydrogen uptake and release [76, 77]. A range of carbon materials have been studied both theoretically (e.g., fullerene, grapheme, etc.) and experimentally (e.g., carbon nanotube, carbon nanofibers, etc.). In transcending from an assortment of porous activated carbons to the intermediary class of MOFs, hydrogen uptake measurements at 77 K had recognized that up to 5 wt% and 7.5 wt% hydrogen can be stored in porous carbons and MOFs, respectively [78]. A preceding disclosure by Chambers et al. which obscured that CNFs could physisorbed up to 67.55 wt% hydrogen, prompted a remarkable renaissance of research related to potential adsorbents [79]. Amongst the diversity of investigations that pursued, some entailed the exploitation of carbon materials using metal dopants while others related to the synthesis of high purity carbons of diverse geometrical structures [58, 80]. Despite so many efforts, none of the carbon nanomaterials could be formed to store hydrogen at a level comparable to the alleged findings of Chambers or the target value specified by the DOE. Similarly, carbon materials had again elicited great expectations after the publicized results of Dillon et al. [58]. In particular, the account of Dillon enthused a significant and apparently somewhat inconsistently-concluding number of studies utilizing single-walled CNTs (SWCNTs), multi-walled CNTs (MWCNTs), CNFs, and carbon nanohorns (CNHs), which had been prepared using chemical vapor deposition (CVD), laser vaporization or arc discharge methods. Nevertheless, more reliable, self-consistent results have been published later with certain carbon nanomaterials [81–83]. A wide variation of uptake values, as found through theoretical calculation and practical experimentations ranging from insignificant to large, is depicted herein for various carbon adsorbents (Table 9.2) [78, 84–114].

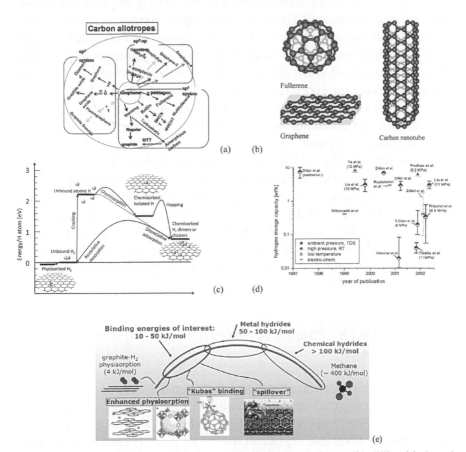

FIGURE 9.5 (a) Allotropes of carbon derived from graphene, (b) Ellipsoid-shaped, tubeshaped, and sheet-shaped carbon nanomaterial, (c) Energy level diagram for the graphenehydrogen system (Reprinted with permission from Tozzini and Pellegrini [72]. © 2013 Royal Society of Chemistry), (d) Year-on-year experimental scatter in hydrogen storage for single-walled carbon nanotubes (Reprinted with permission from Hirscher et al. [96]. © 2003 Elsevier), and (e) Favorable theoretical binding energy regime for easy adsorption and retention of hydrogen (Reprinted with permission from Ströbel et al.[116]. © 2006 Elsevier).

The experimental findings regarding hydrogen storage performance are often less reliable for lack of reproducibility, as judged by many researchers [115–118]. The noticeable fact is that most experimental efforts are visible for CNTs and nanofibers. It appears from Figures 9.5(d) [119] that since end of the last millennium, there is an increasing trend in efforts towards reproducing hydrogen storage results for various carbon nanomaterials, e.g., SWNTs, resulting in considerable scatter in experimental findings. The chief reasons behind such diverse observation are variation of sample quality in terms of purity across different laboratories and variety of processing routes for preparation and characterization of the samples. Among them, Chamber's [78] result concerning gravimetric hydrogen storage is abysmally high (~67 wt%) and that surpasses even promising theoretical estimates too by many folds. The reason behind this observation is attributed to expansion of spacing of graphite layers for excessive pressure generated by high hydrogen pressure and water vapor. However, no other research group till date could claim reproducibility of obtaining such high performance and thus doubt exists about authenticity of purity and characterization of such samples as pointed out by researchers [115, 116]. Most theoretical findings predict hydrogen uptake by carbon lattice to be around 8 wt%, however considering experimental efforts, practical achievable hydrogen adsorption is often less than 1 wt% at ambiance. With lower temperature and higher pressure however, better results could be achieved [100, 106], sometimes even exceeding the theoretical prediction. This is so as for lower temperature and higher pressure the binding energy is more favorable for hydrogen physisorption in pure carbon nanomaterials.

9.5.3   HYDROGEN STORAGE IN FUNCTIONALIZED CARBON NANOMATERIALS

Chemical functionalization of carbon nanomaterial is beneficial in enhancement of hydrogen uptake at similar condition and possibility of retention of hydrogen at higher temperature and lower pressure by bringing the binding energy into the favorable regime of 10–50 kJ mol^{-1} (Figure 9.5(e)) [116]. Fundamentally, this happens due to either extension of the interlayer distance of graphite lattice planes, and an equivalent increase in specific surface area [120, 121] through incorporation of alkali metal atoms or metal halides, or charge transfer from molecular hydrogen (the highest occupied orbital) to empty transition metal orbitals and a subsequent back-donation from the filled d orbitals into the lowest unoccupied orbital of the ligand [122–125], or dissociation of hydrogen into atomic form in presence of catalyst favoring local chemisorption, popularly termed as spillover [126–128]. Significant theoretical and experimental findings concerning hydrogen adsorption in decorated carbon nanomaterials have been shown in Table 9.3 [94, 96, 129–153].

TABLE 9.2 Hydrogen Storage Performance of Pure Carbon Nanomaterials through Physisorption Observed by Researchers Worldwide

Carbon Nanomaterial Type	Storage Condition		Hydrogen Storage Capacity (wt%)	Remarks	Refe-rences
	Temperature	Pressure			
Fullerene	300 K	Ambient	>8	Theoretical; fullerene nanocage	[84]
	300 K	Ambient	~8	Theoretical; charged fullerene	[85]
Graphite	773 K	3.5 MPa	1.6	Experimental	[86]
	300 K	70 MPa	8	Theoretical; fullerene-intercalated graphite	[87]
	77 K	20 bar	1.2	Experimental; exfoliated graphite	[88]
Graphene	300 K	1 MPa	7.4	Experimental	[89]
	Ambient	Ambient	Up to 8	Theoretical; graphene multilayer	[90]
	Ambient	Ambient	7.7	Theoretical	[91]
Activated carbon	293 K	3 MPa	0.39–0.53	Experimental	[92]
Activated carbon fiber	298 K	10 MPa	2.29	Experimental	[93]
Carbon nanotube	Ambient	0.067 MPa	3.5–4.5	Experimental; High purity single-walled	[94]
	Ambient	4.8 MPa	1.2	Experimental; Purified SW	[95]

TABLE 10.2   *(Continued)*

Carbon Nanomaterial Type	Storage Condition		Hydrogen Storage Capacity (wt%)	Remarks	References
	Temperature	Pressure			
	Ambient	0.9 MPa	0	Experimental; Ball-milled SW	[96]
	80 K	~7 MPa	8.25	Experimental; High purity SW	[97]
	295 K	0.1 MPa	0.93	Experimental; Impure SW	[98]
	~300–700 K	Ambient	0.25	Experimental; Purified MW	[99]
	300 K	1.0 MPa	13.8	Experimental; Acid treated MW	[100]
	298 K	3.6 MPa	0.03	Experimental; Low purity MER MW	[101]
	298 K	10 MPa	0.68	Experimental; Random orientation MW	[102]
Nanoporous carbon material	77 K	3.8 MPa	6	Theoretical; quasi-periodic icosahedral	[103]
	Ambient	207 bar	8.13	Experimental; valid for a pore density of 1 cm^3/g and size of 0.9 Å	[104]
Carbon nanofiber	77–300 K	1.5 MPa	1–1.8	Experimental; Purified GNF herringbone	[105]
	77 K	12 MPa	12.38	Experimental; CNF	[106]
	93 K	10 MPa	1	Experimental; CNF	[107]
	298 K	11.35 MPa	53.68	Experimental; GNF platelet	[78]
	298 K	3.6 MPa	<0.1	Experimental; Vapor grown	[101]
Ordered mesoporous carbon	298 K	30 MPa	2.14	Experimental	[108]

TABLE 10.2   *(Continued)*

Carbon Nanomaterial Type	Storage Condition		Hydrogen Storage Capacity (wt%)	Remarks	References
	Temperature	Pressure			
Templated carbon	298 K	10 MPa	1.43	Experimental	[109]
Carbon xerogel	298 K	20	1.0	Experimental	[110]
Carbon sphere	298 K	9	0.25–0.43	Experimental	[111]
Carbon nanosheet	298 K	0.11	0.1–0.26	Experimental	[112]
Carbon cloth	298 K	90 bar	0.5	Experimental	[113]
Carbon nanoscroll	293 K	10 MPa	1.21	Theoretical	[114]

Comparing with Table 9.2, it can be seen that decoration with metal nanoparticle is often capable of causing the adsorption at room temperature and normal pressure feasible which is otherwise almost impossible for pure carbon nanomaterials. A good review highlighting mechanism responsible for enhancement of adsorption by heteroatoms is available [154]. As explained there, heteroatoms work as activation centers to activate hydrogen in a facile manner. To turn that process effective, heteroatoms must be incorporated both geometrically and chemically within carbon network. Efforts in early finding of potential of heteroatoms in tailoring binding energy and desorption temperature of carbon nanomaterials is quite vast, although not entirely exhaustive. A few remarkable recent developments will be highlighted here. A single zirconium atom attached on graphene surface can absorb maximum of 9 H_2 molecules with mean binding energy of 0.34 eV and typical desorption temperature of 433 K leading to a wt% of 11 (Figure 9.6(a)) [155]. Again, calcium atom attached on graphyne surface can absorb up to a maximum of 6 molecular hydrogen (H_2), with a uniform binding energy of ~0.2 eV/H_2, leading to 9.6 wt% of hydrogen (Figure 9.6(b)) [156]. Similar promising results are achieved for CNTs decorated with yttrium (Figure 9.6(c)) [157] and titanium (Figure 9.6(d)) [130]. However, in most cases, experimental observational values are often much less than theoretical estimates. This is because in practical situation it is frequently impossible to reproduce the theoretical conditions due to purity issues, oxidation problems, and atomic clustering of the decorating elements resulting in uneven distribution within the carbonaceous support.

TABLE 9.3 Hydrogen Storage Performance of Functionalized Carbon Nanomaterials Observed by Researchers Worldwide

| Carbon Nanomaterial Type | Storage Condition | | Decorating Element | Hydrogen Storage Capacity (wt%) | Remarks | References |
	Temperature	Pressure				
Fullerene	Ambient	Ambient	Ca	>8.4	Theoretical	[129]
	Ambient	Ambient	Ti	7.5	Theoretical	[130]
Graphene	77 K	Ambient	Zirconia	1.2	Experimental; Reduced GO	[131]
	298 K	4 MPa	Pt and N	4.4	Experimental	[132]
	77 K	10 bar and above	Li	>5	Theoretical; Pillared Graphene Oxide	[133]
	Ambient	Ambient	Ca	~5	Theoretical; zigzag graphene nanoribbon	[134]
	298 K	100 bar	P	2.2	Experimental	[135]
	298 K	3 MPa	Pt	1.4	Experimental	[136]
	298 K	30 bar	Pd3Co	4.6	Experimental; boron-doped	[137]
	Ambient	20 bar	Li	6.5	Theoretical; pillared graphene sheet	[138]
	Ambient	Ambient	Li	12.8	Theoretical	[139]
	Ambient	Ambient	Al	13.79	Theoretical	[140]
Carbon nanotube	473–673 K	0.1 MPa	Li	20	Experimental	[141]
	<313 K	0.1 MPa	K	21	Experimental; Wet H_2	[142]
	473–663	0.1 MPa	Li	0.7–4.2	Experimental; 10–40% pure	[143]

Material	Temperature	Pressure	Dopant	Value	Method	Reference
	Ambient	0.067 MPa	Ti-Al-V	6	Experimental; Sonicated	[94]
	Ambient	0.08 MPa	Fe	<0.005	Experimental; Purified	[96]
	298 K	15.85 atm	Ni	0.812	Experimental; multiwalled	[144]
		12.64 atm	Cu	2.59		
		15.45 atm	Agglomerated Fe	0.909		
	298 K	32 bar	Pd, N	1.25	Experimental; Graphene Nanoplatelets	[145]
	283–335 K	Ambient	Ca	1.05	Experimental; Multiwalled	[146]
			Co	1.51		
			Fe	0.75		
			Ni	0.4		
			Pd	7		
	Ambient	80 bar	B	<2	Experimental	[147]
	Moderate	Ambient	K	4.47	Experimental	[148]
Carbon nanofiber	298	50 bar	Ni	0.56	Experimental decorated CNF grown on ACF substrate	[149]
Porous carbon	298 K	100 bar	P	1.2	Experimental	[150]
Graphyne	Ambient	-	Ca	7	Theoretical	[151]
Graphitic carbon nitride	298 K	4 MPa	Pd	2.6	Experimental	[152]
Superactived carbon	298 K	10 MPa	Pt	1.46	Experimental	[153]
Carbon nanoscroll	293 K	10 MPa	Li	3.78	Theoretical	[114]

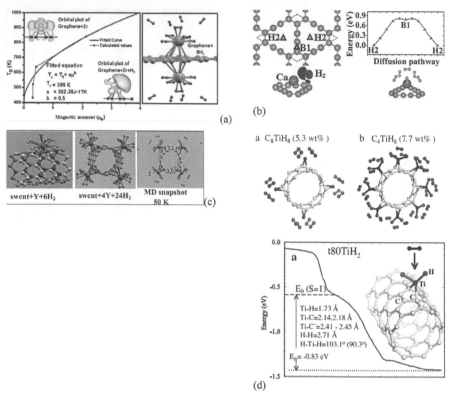

FIGURE 9.6 A quick look of a few theoretical studies explaining role of various dopant atoms in enhancing hydrogen storage capacity of various carbon nanomaterials (a) zirconiumdoped graphene (Reprinted with permission from Yadav et al. [149]. © 2017 American Chemical Society), (b) lithium doped graphyne (Reprinted from Chong et al. [248]. © 2011 American Chemical Society), (c) yttrium-decorated single-walled carbon nanotube (Reprinted with permission from Chakraborty, Modak, and Banerjee [157]. © 2012 American Chemical Society), and (d) titanium-decorated carbon nanotubes (Reprinted with permission from Yildirim and Ciraci [130]. © 2005 American Physical Society.)

9.6   HYDROGEN STORAGE IN POLYMERS

Organic hydrides or liquid organic hydrogen carriers which fix hydrogen by chemical or covalent bonds' formation have been commercialized in existing infrastructure for oil storage [158]. But they fail to fulfill the requisites for being a candidate hydrogen transport material for many safety risks like volatility, flammability, and toxicity. In search of a safer alternative, a few solid-state materials such as metal hydrides [159], porous materials [160] (e.g., activated carbon, metal-organic frameworks (MOFs)) seem to be quite preferable. However, each of them has characteristic limitations over their specific advantages. For example, recovery of hydrogen from metal hydrides is an energy-intensive process involving high temperature. Again, requirement of high-pressure conditions for storing sufficient quantum of hydrogen turns them less attractive. Keeping above in mind, solid formed organic polymers of categories like microporous polymers, hyper-cross-linked polymers, and polymers with intrinsic micro-porosity also find option for in-depth analytical and exploration research for identifying their candidature as efficient hydrogen storage material. This section will focus on efficacy of organic polymers in storing hydrogen in detail. The following subsection would cover a brief and general overview about organic polymers and the other subsequent section would describe progress in research on this frontier including perspectives of decoration with heteroatoms in nanoparticulate form.

9.6.1   ORGANIC POLYMERS: A BRIEF OVERVIEW

A polymer is a substance having a molecular structure composed of essentially from a huge number of collectively bonded similar units [161]. Polymerization is the process by which macromolecules are formed by linking together monomer molecules through chemical reactions [162]. For their organic nature, the polymeric frameworks consist of light elements providing a burden-advantage in many purposes. The long-chain nature sets polymers apart from other materials and gives rise to their characteristic properties. Depending on the nature of the skeletal structure they possess, polymers can be classified as linear, branched, and network or cross-linked polymers [163]. It is the skeletal structure's variation which is responsible for major difference in their properties. A higher crosslink density imparts superior rigidity to a polymer.

Solid organic polymers having visible pores are termed as porous polymers (PP) [164]. They simultaneously contain properties of both a porous material and polymer. They can easily be processed through chemical routes. Some of them may even be dissolved in appropriate solvent and then directly processed using solvent-based methods without destroying their porosity. Hypercross-linked polymers (HCPs) symbolize a class of nano-porous plastics with a broad range of practical and potential applications, e.g., gas sorption and separation, drug delivery, etc., [165]. These polymers manifest numerous significant advantages, including moderate synthetic situations, an enormous stock-room of cheap monomers, robust structures, and good chemical and thermal stabilities. Porous organic polymers (POPs) are those amorphous solids that contain interconnected pores with dispersed sizes and structures, resulting in possessing accessibly large surface areas [166]. Chemical structure of typical microporous polymers of all varieties is shown in Figure 9.7(a) [167]. All the three types of polymers described above are capable of absorbing gas and store within their enclosed intermolecular space at low temperature and high pressure in pure form. The introduction of suitable elements in limited quantity is found to enhance their gas retention properties by many folds at near ambient conditions. The hydrogen adsorption by porous polymer is a completely reversible process and thus it is dependent on applied pressure and temperature. In general, lower is the temperature and higher is the pressure, better is hydrogen adsorption. Measurement of absorption is done both volumetrically (e.g., accelerated surface area and porosimetry system) and gravimetrically (e.g., intelligent gravimetric analyzer). Hydrogen adsorbs only feebly to most surfaces, but adsorption is improved between two surfaces which are close together (i.e., within micropores).

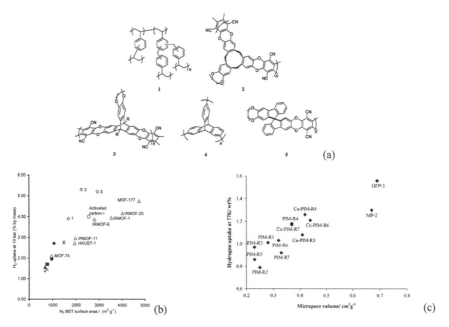

FIGURE 9.7   (a) Typical chemical structure of hypercrosslinked polymer (1), and polymers with intrinsic microporosity (2, 3) (Reprinted with permission from Kato and Nishide [167]. © 2018 Springer Nature) (b) Hydrogen uptake in PIMs in relation with BET surface area and comparison with other porous materials (Reprinted with permission from McKeown, Budd, and Book [187]. © 2007 John Wiley (c) Hydrogen uptake in PIMs in relation with pore volume and comparison with other polymers (Reprinted with permission from Ramimoghadam, Gray, and Webb [188]. © 2016 Elsevier).

9.6.2   HYDROGEN STORAGE IN PURE POLYMERS

In general, porous solid polymers adsorb hydrogen through physisorption, whereas chemisorption is the mechanism by which liquid polymers adsorb hydrogen. But there are a few exceptions. For instance, ketone/alcohol polymer cycles (Table 9.4) [168–174] are capable of storing hydrogen through chemisorption legally and conveniently for designing mobile energy storage device. The diversity of artificial routes for polymers eases the synthesis of numerous porous polymers competent to incorporate multiple chemical functionalities at the pore surface or into the porous framework. In essence, efficacy of hydrogen storage performance through physisorption is vastly dependent on the chemical processing route adopted for their synthesis. Nevertheless, it should be kept in mind that organic microporous polymers are still novices to the variety of potential hydrogen storage materials unlike carbon nanomaterials and only subtle research progress is visible towards maximizing their effectual surface area and optimizing them for efficient uptake and release of hydrogen towards meeting DoE targets (Table 9.1). It is interesting to compare the initial available data for microporous polymers with the best-reported results for other types of material having high surface area. Table 9.5 [175–185] is for comparing hydrogen storage for a number of porous polymers, developed by researchers worldwide. Upon careful observation of the obtained results, the following can be appreciated for hydrogen storage within pure porous polymers.

TABLE 9.4   Hydrogen Storage by Pure Polymers Through Chemisorption

Polymer Name	Characteristic Details	References
Quinaldine-substituted poly(acrylic acid) (PQD) and its hydrogenated 1,2,3,4-tetrahydroquinaldine derivative (PHQD)	Hydrogen release at mild condition (room temperature or ~80°C) at normal pressure	[168, 169]
Fluorenol-fluorenone cycle	Efficient release of hydrogen in presence of iridium catalyst	[170–172]
Polystyrene-poly (cyclohexyl ethylene) cycle	In presence of 10 MPa pressure and palladium catalyst at 200°C	[173, 174]

A high enthalpy of absorption ensures great hydrogen uptake capacity from chemistry point of view. Additionally, a high surface area signifies more and more capacity as adsorption is proportional to surface area [186]. Although hypercrosslinking of nanoporous polymers results in enhancement in hydrogen adsorption capacity, the maximum achievable limit, which is ~ 1.5 wt% still needs upgradation through tailoring of polymer chemistry in order to raise the heat of adsorption for enabling adsorption at higher temperatures [175]. Self-condensation of bischloromethyl monomers leads to generation of a plethora of microporous hypercrosslinked polymers [176]. Networks based on BCMBP display a gravimetric storage capacity of 3.68 wt% at 15 bar and 77.3 K, which is quite appreciable for pure polymers. The porosity in these polymers is governed not only by the level of cross-linking but also by slight changes in the design of the rigid monomer unit. PIMs effectively link the gap between conventional microporous materials and polymers as they possess properties common to both types of material. The prospective offered by the structural variety of PIMs that can be controlled just by the choice of monomer precursors remains still largely unexplored [177]. Available surface area, aggregate pore volume, and distribution of micropores are believed to be the key deciding parameters for their hydrogen storage performance. As shown in Figure 9.7(b), despite having a relatively smaller surface area, many PIMs exhibit superior hydrogen uptake comparing MOFs [187]. Again, though having less pore volume, by virtue of superior pore interconnectivity, many PIMs are better hydrogen retainer than other kinds of polymers (Figure 9.7(c)) [188]. A range of polymers of intrinsic microporosity (PIMs) have been studied by McKeown et al. [166]. The form of hydrogen isotherm is found to be influenced by the micropore distribution. A higher concentration of ultramicropores (pore size <0.7 nm), as observed in PIMs is associated with enhanced low-pressure adsorption. The extended π-electron system and incorporation of sulfur heteroatom did not appear to significantly influence the adsorption of hydrogen for nanoporous polymers with stereo-controlled cores [189]. At ambient temperature, for significantly increasing the heat of adsorptions thus adsorption capacities, new functional groups with sturdy interactions with hydrogen molecules are required. Theoretical studies show that maximum adsorption of hydrogen would happen between two surfaces with a separation of 0.7 nm [190]. This can be achieved by either pore design or inclusion of open metal sites or heteroatoms. These approaches can be easily tailored to microporous polymers, where optimization of characteristics even seems to be much easier. Again, pure organic materials have a discrete weight advantage for being composed solely of light elements. It is conceived from the above studies that improvement of molecular design in terms of rise in molecular weight might lead to rise in hydrogen storage density even beyond DoE target, as exemplified by Kato et al. [167] (Figure 9.8(a)).

TABLE 9.5 Hydrogen Storage Performance of Pure Porous Polymers Observed by Researchers Worldwide

Polymer Name	Type	Storage Condition		Hydrogen Storage Capacity (wt%)	Remarks	References
		Temperature	Pressure			
Amberlite XAD4 poly(styrene-co-divinylbenzene)	Nanoporous	77 K	0.12 MPa	0.8		[175]
Amberlite XAD16 poly(styrene-co-divinylbenzene)				0.6		
Hayesep N poly(divinylbenzene-co-ethylenedimethacrylate)				0.5		
Hayesep B polydivinylbenzene modified with polyethyleneimine				0.5		
Hayesep S poly(divinylbenzene-co-4-vinylpyridine)				0.5		
Wofatit Y77 poly(styrene-co-divinylbenzene)				1.2		
Lewatit EP63 poly(styrene-co-divinylbenzene)				1.3		
Lewatit VP OC 1064 poly(styrene-co-divinylbenzene)				0.7		
Hypersol-Macronet MN200 hypercrosslinked polystyrene	Hypercrosslinked			1.3		
Dichloroxylene (ortho-, meta-, and para-isomers, 0.33 each)	Microporous Hypercrosslinked		1.13 bar	1.3 to 1.56	Process dependent	[176]
4,4¢-Bis(chloromethyl)-1,1¢-biphenyl				1.56		
Bis(chloromethyl)anthracene				1.41		
PIM-1	Soluble non-network PIM		1 bar and 10 bar	0.95 and 1.45		[177]

TABLE 10.5 *(Continued)*

Polymer Name	Type	Storage Condition		Hydrogen Storage Capacity (wt%)	Remarks	References
		Temperature	Pressure			
PIM-7				1.00 and 1.35		[178]
HATN-PIM	Network PIM			1.12 and 1.56		[166]
CTC-PIM				1.35 and 1.70		[178]
Porph-PIM				1.20 and 1.95		[179]
Trip-PIM				1.63 and 2.71		
HCP	Hyper crosslinked			1.27 and 2.75		[180]
PS4AC1	Nanoporous Polymers Containing Stereocontorted Cores	77 K and 298 K	60 bar and 70 bar	2.8 and 0.44		[181]
PT4AC				2.2 0. and 50		
PS4AC2				3.7 and 0.43		
PS4TH				3.6 and 0.45		
poly(glycidyl methacrylate-co-ethylene)	Macroporous copolymer	77 K	0.12 MPa	0.2		[182]
poly(chloromethyl styrene-co-divinylbenzene)	highly polar polymeric sorbents	77 K	0.12 MPa	0.4		[183]
poly(4-vinylpyridine-co-divinylbenzene)				0.6		
Melamine+1,4-diiodobenzene	porous polymer networks of aromatic rings	77 K	1 bar and 20 bar	0.16 and 1.17		[184]
Melamine+ 1,4-dibromobenzene				0.12 and 1.04		
Melamine+ 1,3,5-tribromobenzene				0.10 and 0.94		
diaminobenzene/tribromobenzene	Hypercrosslinked using Buchwald conditions	273 K	9 MPa	0.22		[185]

9.6.3   HYDROGEN STORAGE IN FUNCTIONALIZED POLYMERS

Functional porous polymers can be intended to demonstrate stimuli-responsive uniqueness capable of reversibly altering the pore structure or even toggling between the closed and open porous state after disclosure to environmental stimulus. Such exclusive characteristics are unavailable in other porous materials. Chemical functionalization (in other words, incorporation of heteroatoms) of POP results in improvement in gas storage properties. In essence, incorporation of heteroatoms in a polymer itself is a misnomer as a polymer itself is composed of at least two elemental atoms, i.e., carbon, and hydrogen unlike carbon nanomaterials. Additionally, there is frequent existence of atoms of various non-metals such as oxygen, nitrogen, sulfur, etc. Hence, decoration by heteroatoms typically implies incorporation of either metallic elements or with non-metals having properties more inclined towards metals as per periodic table. Conjugated microporous polymers (CMPs) [191–194], covalent organic frameworks (COFs) [195–199], covalent triazine frameworks (CTFs) [200–202], porous aromatic frameworks (PAFs) [203–205],metal-organic frameworks (MOF) are a few among these highly assorted varieties where substantial mixing of foreign elements results in change in entire nature of the mother polymer and generation of new variety of polymers typically known as coordination polymers. Though these are fairly well performer for hydrogen storing, typical weight advantage of pure polymers is somehow compromised for incorporation of heteroatoms in large quantities. Discussion within this section thus will be restricted to experimental and theoretical progress investigating hydrogen storage performance of POPs functionalized marginally with elemental nano-particles.

Most of such efforts are unfortunately found to be theoretical in nature and there is rarity of experimental instances in this direction till date. Decorating elements such as Ti, Li, Sc etc. are found to cause sufficient shift of binding energy and desorption temperature towards favorable condition to enable porous polymers to contain enough amount of hydrogen near ambient temperature and pressure. For example, Lee et al. showed through first principle calculation that Ti-decorated *cis*-polyacetylene might attain a reversibly usable gravimetric density of 7.6 wt% [206] (Figure 9.8(b)). A similar study involving ethylene by Durgun et al. concludes that Ti is capable of raising hydrogen intake capacity of ethylene complex up to 14 wt% [207]. For CMPs, Li doping is found to enhance hydrogen uptake up to 6.1 wt% at 77 K and normal pressure [208] (Figure 9.8(c)). Sc and Li functionalized [2,2] paracyclophane complexes are found to hold up to 10

H_2 molecules and 8 H_2 molecules by charge polarization mechanism with hydrogen weight percentage of 11.4 and 13.5, respectively [209]. Such instances are promising and encouraging for endeavoring future experiments for validating such theoretical findings and ruling out any oversimplifying assumption during theoretical recalculation for accomplishing more realistic observation. Boron was experimentally found to enhance hydrogen storage property of POP in the form of carborane by Yuan et al. [210]. They found that boron is efficient in raising initial isosteric heats of adsorption toward hydrogen by 3–5 kJ/mol. Pd/MIL-101 Nanocomposites have been recently found to exhibit reducing hydrogen absorption enthalpy by Zhao et al. [211]. The above findings strongly suggest that decorated organic porous polymer has huge potential for cost-effective exploitation as a future hydrogen carrier.

9.7   THEORETICAL INSIGHTS FOR HYDROGEN STORAGE

9.7.1   IMPORTANCE OF THEORETICAL SIMULATIONS FOR HYDROGEN STORAGE

Theoretical simulations play an important role in predicting the interaction mechanism of hydrogen molecules with the nanomaterials as well as with the functionalized atoms or groups in the case of functionalized Nanomaterials. Simulation techniques can give important information like preferred adsorption configuration, adsorption energy, charge transfer, etc. which are difficult to get from experiments. With the invention of high speed, large memory supercomputers, extensive computer simulations are now very much possible. Also, so many well-established computer codes, both open access and commercial are available to compute theoretical data on hydrogen storage. So, the availability of trusted software packages and supercomputers makes it possible to generate hydrogen storage database for a large number of systems and recommend the experimentalist the potential systems to try in experiments. Theoretical simulations can provide nice picture on orbital interactions and charge transfer which is difficult or may not be possible to get from experimental measurements. From the simulations data, it is straightforward to conclude whether specific H_2 is having physisorption, chemisorptions, or Kubas [122, 212] types of interactions. The stability of the structures and desorption temperature can also be obtained from theoretical simulations. One of the important issues for hydrogen storage in metal-doped carbon nanomaterial or polymer is the metal-metal cluster.

From the computation of diffusion energy barrier of metals, we can get idea on possible metal loading pattern to avoid metal-metal clustering. But there are challenges and issues for predicting hydrogen wt% from theoretical simulations. Simulations results are sensitive to the simulations methods and setting parameters. For example, LDA exchange-correlation [213] over-binds the H_2 molecules whereas GGA exchange-correlation [214] under-binds the H_2 molecules. So, we need to be careful regarding choice of the simulation methods and prediction coming from simulation details. For the adsorption of H_2 molecules on carbon nanostructures or polymers, weak Van der Waals interactions are present. So, we should carefully include those interactions while computing binding energy of H_2 molecules on the host. Keeping in mind, the role of theoretical simulations for predicting hydrogen storage parameters and the limitation of simulations methods, the following sub-sections will highlight overview of quantum simulations, important parameters obtained from theoretical data, importance of dispersion corrections and sensitivity of theoretical data.

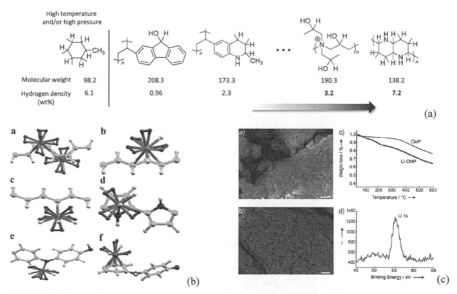

FIGURE 9.8 (a) Progressive improvement in hydrogen storage density through chemical modification in polymer (Reprinted with permission from Kato and Nishide [167. © 2018 Springer Nature), (b) Atomic structures of Ti decorated polymers with maximum no. of hydrogen molecules attached with Ti atom (Reprinted with permission Lee and Choi [180]. © 2006 American Physical Society), (c) Existence of Li in ionic state with one negative charge per Li atom within Li-doped CMP polymer (Reprinted with permission Li et al. [208]. © 2010 John Wiley).

9.7.2   OVERVIEW OF QUANTUM SIMULATIONS

Quantum simulation is very powerful in terms of providing the physical insight of the system as well as producing data which can be compared from experimental data. Quantum simulations find applications in many branches of science and engineering. In last few decades, it has gain popularity due to development of high speed, large memory supercomputer. One of the notable features of quantum simulations is that it does not require any model potential to be provided by the user. Quantum simulations involve solving the Schrodinger's equation for the concerned system containing so many numbers of electrons and ions. The major challenge for quantum simulations is that the equations like the Schrodinger's equation are difficult to solve analytically except for a few simple systems. Solving such an equation needs sophisticated numerical techniques and many approximations. These approximations and sophisticated numerical techniques have given birth to various quantum simulations methods, e.g., density functional theory (DFT) [215–217], Hartree-Fock [218, 219], Quantum Monte Carlo [220–222], coupled cluster [223, 224], multi-reference configuration interaction [225], etc. Each of the methods has their own merits and demerits.

9.7.2.1   MANY-BODY HAMILTONIAN

The central idea of quantum mechanical simulations is to find the electronic structure of the concerned system by solving the many-body Schrödinger equation. For a system containing the positively charged nuclei and negatively charged electrons, the non-relativistic many-particle Hamiltonian as:

$$\hat{H} = \hat{T}_{nucl} + \hat{T}_{elec} + \hat{V}_{n-e} + \hat{V}_{e-e} + \hat{V}_{n-n} \tag{9.1}$$

where the different terms are as follows:

$$\hat{T}_{nucl} = -\frac{\hbar^2}{2}\sum_i \frac{\nabla^2_{R_i}}{M_i}, \hat{T}_{elec} = -\frac{\hbar^2}{2}\sum_i \frac{\nabla^2_{r_i}}{m_i},$$

$$\text{and}\, \hat{V}_{n-e} = -\frac{1}{4\pi\varepsilon_0}\sum_{i,j} \frac{Z_i e^2}{\left|R_i - r_j\right|}, \hat{V}_{e-e} = \frac{1}{8\pi\varepsilon_0}\sum_{i\neq j} \frac{e^2}{\left|r_i - r_j\right|}, \hat{V}_{n-n} = \frac{1}{8\pi\varepsilon_0}\sum_{i\neq j} \frac{e^2 Z_i Z_j}{\left|R_i - R_j\right|}$$

$$\tag{9.2}$$

First-term represents the kinetic energy operators for the nuclei, second term is the kinetic energy operators for the electrons, third, fourth, and fifth terms indicate the Coulomb interactions between nuclei-electron, electron-electron, and nuclei-nuclei respectively. M_i and m_i refer to the masses of the nuclei and electrons at the positions R_i and r_i, respectively. The wave functions ($\psi\{R_n, r_e\}$) of the system can be found out from the above Hamiltonian by solving the following Schrödinger equation:

$$\hat{H}\psi\{R_n, r_e\} = E\psi\{R_n, r_e\} \tag{9.3}$$

As it is very difficult to solve it, we first use Born-Oppenheimer approximation to separate the electron and nuclei degree of motion. Since the velocities of the nuclei, are much less than that of the electrons, then nuclei can be considered to be fixed at their position. So, the nuclei act as external particles with a potential and the electrons move [226]. With this assumption, the kinetic energy part of the Hamiltonian can be neglected and the Coulomb interactions between the nuclei can be taken as constant. So, the Hamiltonian of much electron system is reduced to:

$$\hat{H} = \hat{T}_{elec} + \hat{V}_{ext} + \hat{V}_{e-e} \tag{9.4}$$

But still solving the Schrödinger equation with the reduced Hamiltonian is quite challenging and different methods have been formulated. Now we will mention some of the methods for quantum simulations.

9.7.2.2   HARTREE-FOCK METHOD

Hartree-Fock method [218, 219] is the corrected version of the Hartree method where wave function of many-electron systems is considered as a simple product of single-electron orbitals. But as the Hartree method does not satisfy the Pauli's exclusion principle, in Hartree-Fock method, the wave function is described by the anti-symmetries many-electron wave function using the slater determinant.

$$\psi_{HF}(x_1, x_2, \cdots, x_N) = \frac{1}{\sqrt{N!}} \begin{bmatrix} \varphi_1(x_1) & \cdots & \varphi_1(x_N) \\ \vdots & \ddots & \vdots \\ \varphi_N(x_1) & \cdots & \varphi_N(x_N) \end{bmatrix} \tag{9.5}$$

This leads to the Hartree-Fock equation given by:

$$\varepsilon_i \varphi_i(\vec{r})=[\hat{T}_{elec}+\hat{V}_{ext}+\hat{V}_{e-e}]\varphi_i(\vec{r})-\frac{1}{2}\sum_j \int \varphi_j^*(\vec{r'})\varphi_i(\vec{r'})\frac{1}{|\vec{r}-\vec{r'}|}\varphi_i(\vec{r})d^3\vec{r'} \quad (9.6)$$

The main demerit of this method is that the exchange term is non-local and it neglects the electron correlations. Hartree-Fock method gives good results for molecules but fails for metals. This scheme is backbone for various Post Hartree-Fock methods like Coupled cluster [223, 224], Multi-reference Configuration Interaction [225], etc. where correlation effects are taken care of.

9.7.2.3   DENSITY FUNCTIONAL THEORY (DFT)

Density functional theory (DFT) is one of the powerful and fast quantum simulation methods which are widely used in many branches of science and engineering. DFT is based on two theorems given by Hohenberg and Kohn [227] in 1964 and followed by few equations formulated by Kohn and Sham [228]. Two theorems of Hohenberg and Kohn are as follows:

1. **First Theorem:** There is a one-to-one correspondence between the ground-state density $\rho(\vec{r})$ of a many-electron system and the external potential $\hat{V}_{ext}$. An immediate consequence is that the ground-state expectation value of any observable $\hat{O}$ is a unique functional of the exact ground-state electron density.

$$\langle \psi|\hat{O}|\psi \rangle = O[\rho] \quad (9.7)$$

2.   **Second Theorem:** For $\hat{O}$ being the Hamiltonian $\hat{H}$, the ground-state total energy functional:

$$H[\rho] \equiv E_{V_{ext}}[\rho] \quad (9.8)$$

is of the form:

$$E_{V_{ext}}[\rho] = \underbrace{\left\langle \psi | \hat{T} + \hat{V} | \psi \right\rangle}_{F_{HK[\rho]}} + \left\langle \psi | \hat{V}_{ext} | \psi \right\rangle = F_{HK[\rho]} + \int \rho(\bar{r}) V_{ext}(\bar{r}) d\bar{r}$$

(9.9)

where, the Hohenberg-Kohn density functional $F_{HK[\rho]}$ is universal for any many-electron system. $E_{V_{ext}}[\rho]$ reaches its minimal value (equal to the ground-state total energy) for the ground state density corresponding to $\hat{V}_{ext}$.

The theorem tells the existence of a one-to-one correspondence between the external potential and the electron density. So, everything regarding the system can be known from the functional F_{HK}. By minimizing the functional F_{HK}, we can find out the exact energy and density of the many electron system. But the problem is that F_{HK} is not known which makes it difficult for the implementation of this formulism for real system.

9.7.2.4 THE KOHN-SHAM EQUATIONS

Kohn-Sham equations led the foundation for the practical implantation of DFT. They have replaced the real interacting system by a non-interacting system with some external potential. The Hamiltonian is now given by:

$$\hat{H}_{KS} = \hat{T}_0 + \hat{V}_H + \hat{V}_{xc} + \hat{V}_{ext}$$

(9.10)

With $\hat{T}_0$ being the kinetic energy operator and $\hat{V}_H$, $\hat{V}_{xc}$ and $\hat{V}_{ext}$ are the Hartree, exchange-correlation, and external potentials respectively. Now the energy functional can be expressed as:

$$E_{V_{ext}}[\rho] = \underbrace{T_0[\rho] + E_H[\rho] + E_{xc}[\rho]}_{F_{HK[\rho]}} + \int \rho(\bar{r}) V_{ext}(\bar{r}) d\bar{r}$$

(9.11)

With $F_{HK[\rho]}$ denoting the sum of the first three terms and $T_0[\rho]$ being the kinetic energy functional along with $E_H[\rho]$ and $E_{xc}[\rho]$ representing the Hartree and exchange-correlation functional. Now the ground state density can be obtained from the solution of Kohn-Sham equation given by:

$$\left[-\frac{\hbar^2}{2m_e}\nabla+\hat{V}_{eff}\left(\vec{r}\right)\right]\psi_i(\vec{r})=\varepsilon_i\psi_i(\vec{r}) \qquad (9.12)$$

with

$$\hat{V}_{eff}(\vec{r})=\hat{V}_H+\hat{V}_{xc}+\hat{V}_{ext} \qquad (9.13)$$

and $\hat{V}_{xc}$ and $\hat{V}_H$ are given by:

$$\hat{V}_{xc}=\frac{\delta E_{xc}[\rho]}{'n},\hat{V}_H=\frac{e^2}{4\pi\varepsilon_0}\int\frac{\rho(\vec{r}')}{\left|\vec{r}-\vec{r}'\right|}d\vec{r}' \qquad (9.14)$$

We can see that the only unknown term is exchange-correlation functional. The above equation can be solved using a self-consistent procedure. So the accuracy of DFT data depends on how good the exchange-correlation functional can be formulated. LDA (local density approximation) [213] and GGA (generalized gradient approximation) [214] are two widely used exchange-correlation functional. LDA is based on homogenous electron gas model and is expressed as [214]:

$$E_{xc}^{LDA}[\rho]=\int\rho(\vec{r})\varepsilon_{xc}(\rho(\vec{r}))d\vec{r} \qquad (9.15)$$

where, the exchange part of the ε_{xc} can be obtained analytically and the correlation is obtained numerically using quantum Monte-Carlo simulations [213].

In GGA, the gradient of the density is also taken into account. There exist different versions of GGA exchange-correlation Functional, e.g., Perdew-Burke-Ernzerhof (PBE) [214], and Becke-Lee-Yang-Parr (BLYP) [229, 230]. The notable difference between GGA and LDA results that LDA overbinds the system whereas GGA underbinds the system.

To perform simulations with different quantum mechanical methods, various software packages are available. Softwares like Gamess [231] or Gaussian [232] are designed to deal with molecular structures using a localized basis set like the Gaussian center orbitals or atomic orbitals. Codes like VASP [233] or Quantum-espresso [234] which are designed to deal with metals and other solids aiming for band structure calculations use the plane wave basis set. Packages like WIEN2K [235] or ELK [236] are made mainly to deal with heavy elements where the pseudopotential methods are not suitable and all the electron potential should be considered.

9.7.3 PROPERTIES FROM THEORETICAL SIMULATIONS

Theoretical simulations can provide very beautiful and clear picture regarding adsorption configuration, adsorption energy, charge transfer, and orbital interactions which are difficult to get in experiments.

9.7.3.1 ADSORPTION SITE

Finding the adsorption configuration with minimum energy is an important step for hydrogen storage simulations. We need to perform geometry optimization for a longer period till the forces between the atbecomeomes very less (typically 0.01 eV/Å). For the hydrogen adsorption on carbon Nanomaterials as well as polymer due to involvement of weak Van der Waal's forces, the convergence may take longer time. Sometimes, the gas molecules may get stuck to the local minima and we may miss the actual configuration with minimum energy. To avoid this, it is suggested to make 2–3 different initial configurations and allow them relax separately and choose the one with minimum energy as the stable configuration. If the system contains transition metal with magnetic signature, the relaxation and total energy calculation should be performed with spin polarization into consideration. The stable configuration also depends on simulations setting parameters such as energy cut off for the plane wave as well as K-points for sampling the Brillouin zone. So, one needs to test the convergence with respect to energy cut off for the plane wave as well as number of K-points. The distance of the hydrogen molecule with the nearest atom of the host or the dopant can give idea regarding strength of the bonding. Lower the distance (bond length) stronger is the adsorption. Figure 9.9 displays two possible adsorption configurations of Y on graphyne [237], one on hexagonal site and other on triangular site. The adsorption is strong on triangular site with binding energy of -4.05 eV compared to hexagonal site with –3.15 eV. Figure 9.9 also depicts the configuration of 9 H_2 attached on hexagonal site and 5H_2 attached on triangular site.

9.7.3.2 ADSORPTION ENERGY

The adsorption energy of H_2 is a key parameter in finding the capability of hydrogen storage of the material. For practical hydrogen storage device, this energy must fall within a window. If the adsorption energy is too high, it will be difficult to release them, whereas if the H_2 molecules are attached weakly, then they may come out of the system before using. In this regard, DoE [45] has prescribed that the binding energy of H_2 must fall within the range 0.2–0.7 eV for practical implementation of hydrogen as fuel. The adsorption energy of H_2 on the host can be computed from the expression:

$$E_{ads} = E_{Host+H_2} - E_{Host} - E_{H_2}$$ (9.16)

where, E_{Host+H_2} is the energy of the host (nanostructure + dopant) plus H_2, E_{Host} is the energy of the host and E_{H_2} is the energy of an isolated H_2 molecule. As per the convention, negative value of adsorption energy indicates that binding is energetically favorable. All three energies in the above equation should be computed with same simulations setting as the computed energy may vary with the setting parameters. The adsorption energy may be different for different adsorption sites. If there is no periodic boundary condition in any direction (for molecules, clusters, etc.) sufficient vacuum should be considered to avoid interactions between periodic images. Table 9.6 presents the binding energy of the metal Y and H_2 molecules on two different sites of on graphyne [237]. We can notice that binding energy is given with GGA and GGA+ Van der Waal's corrections. With the inclusion of Van der Waal's corrections, there is slight increase in binding energy of H_2 molecules.

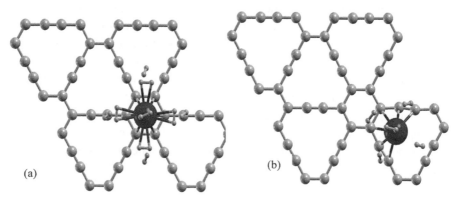

(a)   (b)

FIGURE 9.9   Two possible adsorption sites of Y on graphyne (a) Hexagonal site with metal binding energy of –3.15 eV; the configuration of 9 H_2 molecules it can absorb and (b) Triangular site with metal binding energy of –4.05 eV; the configuration of 5 H_2 molecules it can absorb (Reprinted with permission f Gangan et al. [237]. © 2019 Elsevier).

TABLE 9.6 The Computed Binding Energy per H2 and Desorption Temperature for All Configurations for Y Doped Graphyne Using GGA and GGA + Van Der Waals Corrections

Configurations	Binding Energy (eV)				Average Desorption Temperature (K)			
	Site 1		Site 2		Site 1		Site 2	
	GGA	GGA vDW	GGA	GGA vDW	GGA	GGA vDW	GGA	GGA vDW
Y	3.15	3.433	4.05	4.32	–	–	–	–
H$_2$	0.304	0.357	0.29	0.35	388	456	371	447
3H$_2$	0.292	0.378	0.30	0.40	373	483	383	511
5H$_2$	0.28	0.376	0.26	0.35	356	480	332	447
9H$_2$	0.261	0.3	–	–	333	383	–	–
Average BE	0.297	0.366	0.28	0.366	379	468	358	468

Source: Reprinted with permission from Gangan et al. [237], © 2019 Elsevier.

9.7.3.3   CHARGE TRANSFER AND ORBITAL INTERACTIONS

Theoretical insight regarding bonding and charge transfer mechanism can be obtained from simulation data. We can get quantitatively the amount of charge getting transferred between the bonded species. From the charge transfer we can get some idea regarding bonding mechanism. The H$_2$ molecule can be attached on the host through physisorption, chemisorptions, or Kubas types of interaction. The bonding is strong for the chemisorption, weak for physisorption and intermediate for the Kubas types of interaction. More the charge transfer, stronger is the bonding. For pure polymer or the carbon Nanomaterials, the adsorption is physisorption and the charge transfer is negligible. For chemisorptions, the hydrogen atoms get dissociated and charge transfer is large. Both physisorption and chemisorptions are not suitable for practical implementation of hydrogen storage. For metal doped carbon nanomaterials or polymer, there is charge transfer from metal d orbitals to H's orbital followed by back donation through Kubas interaction. In this case, the binding energy comes in the range 0.2–0.7 eV as prescribed by DoE and is most suitable for practical implementation of hydrogen storage. So, the charge transfer can give a picture whether the interaction is chemisorptions, physisorption, or Kubas types of interactions [122, 212]. To find the charge transfer one need to take care regarding number of valence electrons taken for the simulations. Bader charge analysis [238] and Mulliken's charge analysis [239] are two powerful techniques to find the amount of charge transfer during the orbital interactions. To visualize the spatial variation of charge density, the pictorial plot of charge density can be obtained using visualize software like XCRYSDEN [240], VESTA [241] etc. The orbital interactions and hybridization can be obtained qualitatively from plot of partial density of states which describe the contribution of Density of states from each orbital of each atom. There should consistency between the charge transfer from Bader charge analysis, analysis of partial density of states and charge density plot. Chakraborty et al. [157] have shown that when Y atom is absorbed on single-walled carbon nanotube (SWCNT), there is 1.1446e charge transfer from Y atom to SWCNT. Figure 9.10 depicts [157] the charge density plot for the charge density difference between SWCNT+Y and SWCNT for the isovalue of 0.2e and -0.2e. We can notice that there is a polarization between inner and outer surfaces due to charge transfer and charge redistribution.

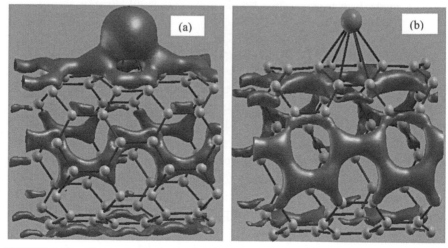

FIGURE 9.10   Charge density isosurface for charge density difference between Y-SWCNT and SWCNT systems (a) iso-value 0.2e (b) iso-value –0.2e (color online). (Reprinted with permission from Chakraborty, Modak, and Banerjee [157]. © 2012 American Chemical Society).

9.7.4   IMPORTANCE OF DISPERSION CORRECTIONS FOR HYDROGEN STORAGE

As the H_2 adsorption involves weak Van der Waals interactions, we must include dispersion corrections in our simulations as the standard DFT with GGA exchange-correlation functional can't handle long term electron correlation effects. The dispersion correction increases the binding energy of H_2 molecules slightly. There exist various schemes for incorporating the dispersion corrections in DFT and the binding energy of H_2 molecules will differ little bit for various dispersion schemes. Here we will briefly outline the different schemes employed for dispersion corrections.

1. **Grimme's DFT-D2** [242]: In this scheme, the total energy is expressed as a sum of the Kohn–Sham energy, E_{DFT}, and vdW semi-empirical pair correction, $E_{disp}^{(2)}$. The pair corrections depend on two interacting atoms and independent of environment.
2. **Grimme's DFT-D3:** Here the environment dependent dispersion coefficients (A three body component) were considered during the DFT simulations to improve the accuracy of Grimme's DFT-D2 scheme.
3. **vdW-DF** [244–246]: In this approach, dispersion interactions are described in terms of electron density and no additional inputs needed.
4. **optPBE-vdW:** It is a different version of vdW-DF.

In Table 9.7, we have presented the adsorption energy of Co molecule on graphene and defected graphene employing different schemes of dispersion corrections [247]. It is clear from Table 9.7 that various scheme yields different adsorption energy and we get a window for adsorption energy. We can notice that the PBE+vdW and DFT-D3 give almost identical binding cnergies, 100 meV higher than the DFT-D2 method. So, we need to select the proper dispersion correction schemes depending upon the system used for the simulations.

TABLE 9.7 Comparison of the Calculated Adsorption Energy for Adsorption of CO Molecule on Graphene and Defective Graphene for Different vdW Functional

Method	Graphene		Defected Graphene	
	Adsorption Energy (eV)	Distance (Å)	Adsorption Energy (eV)	Distance (Å)
PBE + vdW	−0.11	2.88	−2.15	1.33
DFT-D2	−0.09	2.87	−2.25	1.33
DFT-D3	−0.10	2.92	−2.17	1.33
vdW-DF	−0.14	3.02	−1.80	1.33
optB88-vdW	−0.11	2.77	−2.04	1.33
optPBE-vdW	−0.16	2.93	−2.01	1.33

Source: Jiang et al. [247],with permission.

9.7.5 SENSITIVITY OF THEORETICAL DATA

The data obtained from DFT simulations depend on the exchange-correlation functional, simulations parameters, dispersion correction schemes and simulation software. So, one need to be careful regarding choice of the simulation parameters. As the binding energy is computed from difference of energy, simulations should be considered with identical setting parameters otherwise, it may give wrong result. Here we will mention two important parameters, namely, energy cut-off and k-point grid.

1. **Energy Cut-Off:** It is a key parameter to get accurate value of ground state properties. Before taking the data, one must check the convergence of energy with respect to cut-off energy. For a periodic system, the solution of the Schrödinger equation can be expressed as [214]:

$$\phi_n(r) = \exp(ik \cdot r) f_n(r) \qquad (9.17)$$

where, the periodic function $f_n(r)$ can be expanded in terms of finite number of plane waves with the reciprocal lattice vectors of the crystal:

$$f_n(r) = \Sigma_G C_{n,G} \exp[iG \cdot r] \qquad (9.18)$$

The summation is carried out over the reciprocal lattice vector G. Now the wave function can be written as:

$$\phi_n(r) = \Sigma_G C_{n,k+G} \exp[i(k+G) \cdot r] \qquad (9.19)$$

where, $C_{n,k+G}$ indicates the coefficients of the plane waves with kinetic energy:

$$E = \frac{h^2}{2m} |k+G|^2 \qquad (9.20)$$

This gives a kinetic energy cut-off for the plane waves given by:

$$E_{cut} = \frac{h^2}{2m} G_{cut}^2 \qquad (9.21)$$

and the infinite sum becomes:

$$\phi_i(r) = \sum_{|G+k|<G_{cut}} C_{i,k+G} \exp[i(k+G)\cdot r] \qquad (9.22)$$

Higher the energy cut-off, expensive is the simulation and accurate is the result. One should take the energy cut-off such that energy convergence is achieved.

2. **k-Point Grid:** It is another key parameter for DFT simulations which is used for sampling the Brillouin-zone. One needs to check the convergence of energy with respect to K-points. As the K-points are in the reciprocal space, higher the cell length in any direction, lower should be the number of K-points in that direction. K-points are required only in the direction where periodicity is there. The computational cost increases with increase in K-points.

9.8 CONCLUSIVE REMARKS WITH FUTURE SCOPES

In this chapter, we have presented an overview of various materials for efficient hydrogen storage. We have also included the features and capabilities of quantum simulations techniques in predicting various hydrogen storage parameters, e.g., binding energy, hydrogen wt%, desorption temperature, etc. Functionalized carbon Nanomaterials are future for efficient hydrogen storage at ambient condition. This field has been explored quite a lot in theoretical prediction. But the experimental results are so far not very promising and experimental wt% in most cases is lower than theoretical prediction. The main issue is to synthesize functionalized carbon Nanomaterials and polymer. During the synthesis, oxygen interference, metal-metal clustering, impurity, defects, etc. hinders the hydrogen storage and experimentally obtained wt% comes much below the requirement set by DoE. So, the developing sophisticated and innovated method for synthesis of functionalized carbon Nanomaterials and polymer is the key for implementing hydrogen fuel in transport applications.

ACKNOWLEDGMENTS

BC would like to thank Dr. R. C. Rannot and Dr. A. K. Mohanty for their support and encouragement. BC would also like thank Dr. S. Banerjee for inspiration and scientific discussions. GS would like to thank Dr. R. N Singh Dr. V. Kain and Dr. Madangopal Krishnan for their support.

KEYWORDS

- **carbon nanomaterials**
- **conjugated microporous polymers**
- **density functional theory**
- **hydrogen fuel**
- **hydrogen storage**
- **organic polymers**

REFERENCES

1. Shafiee, S., & Topal, E., (2009). When will fossil fuel reserves be diminished? *Energy Policy, 37*(1), 181–189.
2. Hoel, M., & Kverndokk, S., (1996). Depletion of fossil fuels and the impacts of global warming. *Resource and Energy Economics, 18*(2), 115–136.
3. Mazloomi, K., & Gomes, C., (2012). Hydrogen as an energy carrier: Prospects and challenges. *Renewable and Sustainable Energy Reviews, 16*(5), 3024–3033.
4. Bechtold, R., (1997). *Alternative Fuels Guidebook.* SAE.
5. National Research Council, (2013). *Transitions to Alternative Vehicles and Fuels.* National Academies Press.
6. Lee, S., Speight, J. G., & Loyalka, S. K., (2014). *Handbook of Alternative Fuel Technologies.* CRC Press.
7. National Research Council, (2008). *Transitions to Alternative Transportation Technologies: A Focus on Hydrogen.* National Academies Press.
8. Singh, S., Jain, S., Venkateswaran, P. S., Tiwari, A. K., Nouni, M. R., Pandey, J. K., & Goel, S., (2015). Hydrogen: A sustainable fuel for future of the transport sector. *Renewable and Sustainable Energy Reviews, 51*, 623–633.
9. Mandal, T. K., & Gregory, D. H., (2010). Hydrogen: A future energy vector for sustainable development. *Proceedings of the Institution of Mechanical Engineers, Part C: Journal of Mechanical Engineering Science, 224*(3), 539–558.
10. Yartys, V. A., & Lototsky, M. V., (2004). An overview of hydrogen storage methods. In: *Hydrogen Materials Science and Chemistry of Carbon Nanomaterials* (pp. 75–104). Springer, Dordrecht.
11. Walker, G., (2008). *Solid-State Hydrogen Storage: Materials and Chemistry.* Wood head Publishing, Elsevier.
12. Stetson, N. T., (2012). Hydrogen storage overview. In: *DoE Annual Merit Review and Peer Evaluation Meeting.*

13. Durbin, D. J., & Malardier-Jugroot, C., (2013). Review of hydrogen storage techniques for on board vehicle applications. *International Journal of Hydrogen Energy, 38*(34), 14595–14617.

14. Ni, M., (2006). An overview of hydrogen storage technologies. *Energy Exploration and Exploitation, 24*(3), 197–209.

15. Beaudin, M., Zareipour, H., Schellenberglabe, A., & Rosehart, W., (2010). Energy storage for mitigating the variability of renewable electricity sources: An updated review. *Energy for Sustainable Development, 14*(4), 302–314.

16. Abe, J. O., Ajenifuja, E., & Popoola, O. M., (2019). Hydrogen energy, economy, and storage: Review and recommendation. *International Journal of Hydrogen Energy.*

17. Chen, H., Cong, T. N., Yang, W., Tan, C., Li, Y., & Ding, Y., (2009). Progress in electrical energy storage system: A critical review. *Progress in Natural Science, 19*(3), 291–312.

18. Dahiya, R. P., (1985). Progress in hydrogen energy. In: *Proceedings of the National Workshop on Hydrogen Energy* (pp. 1019–1036). New Delhi.

19. Züttel, A., Remhof, A., Borgschulte, A., & Friedrichs, O., (2010). Hydrogen: The future energy carrier. *Philosophical Transactions of the Royal Society A: Mathematical, Physical, and Engineering Sciences, 368*(1923), 3329–3342.

20. Winter, C. J., & Nitsch, J., (2012). *Hydrogen as an Energy Carrier: Technologies, Systems, Economy.* Springer Science and Business Media.

21. https://www.energy.gov/eere/fuelcells/h2scale (accessed on 10 August 2020).

22. Fayaz, H., Saidur, R., Razali, N., Anuar, F. S., Saleman, A. R., & Islam, M. R., (2012). An overview of hydrogen as a vehicle fuel. *Renewable and Sustainable Energy Reviews, 16*(8), 5511–5528.

23. Veziro, T. N., & Barbir, F., (1992). Hydrogen: The wonder fuel. *International Journal of Hydrogen Energy, 17*(6), 391–404.

24. Pudukudy, M., Yaakob, Z., Mohammad, M., Narayanan, B., & Sopian, K., (2014). Renewable hydrogen economy in Asia-opportunities and challenges: An overview. *Renewable and Sustainable Energy Reviews, 30*, 743–757.

25. Crabtree, G. W., Dresselhaus, M. S., & Buchanan, M. V., (2004). The hydrogen economy. *Physics Today, 57*(12), 39–44.

26. Edwards, P. P., Kuznetsov, V. L., David, W. I., & Brandon, N. P., (2008). Hydrogen and fuel cells: Towards a sustainable energy future. *Energy Policy, 36*(12), 4356–4362.

27. Prasath, B. R., Leelakrishnan, E., Lokesh, N., Suriyan, H., Prakash, E. G., & Ahmed, K. O., (2012). Hydrogen operated internal combustion engines-a new generation fuel. *International Journal of Emerging Technology and Advanced Engineering, 2*(4), 52–57.

28. https://www.slideshare.net/SridharSibi/hydrogen-fuel-amp-its-sustainable-development (accessed on 3 August 2020).

29. Williams, Q., & Hemley, R. J., (2001). Hydrogen in the deep earth. *Annual Review of Earth and Planetary Sciences, 29*(1), 365–418.

30. Ohta, T., (2013). *Solar-Hydrogen Energy Systems: An Authoritative Review of Water-Splitting Systems by Solar Beam and Solar Heat: Hydrogen Production, Storage, and Utilization.* Elsevier.

31. Balat, H., & Kırtay, E., (2010). Hydrogen from biomass-present scenario and future prospects. *International Journal of Hydrogen Energy, 35*(14), 7416–7426.

32. Iwahara, H., Esaka, T., Uchida, H., & Maeda, N., (1981). Proton conduction in sintered oxides and its application to steam electrolysis for hydrogen production. *Solid State Ionics, 3*, 359–363.

33. Cortright, R. D., Davda, R. R., & Dumesic, J. A., (2011). Hydrogen from catalytic reforming of biomass-derived hydrocarbons in liquid water. In: *Materials for Sustainable Energy: A Collection of Peer-Reviewed Research and Review Articles from Nature Publishing Group,* (pp. 289–292).

34. Turner, J. A., (2004). Sustainable hydrogen production. *Science, 305*(5686), 972–974.

35. Holladay, J. D., Hu, J., King, D. L., & Wang, Y., (2009). An overview of hydrogen production technologies. *Catalysis Today, 139*(4), 244–260.

36. Ni, M., Leung, M. K., Leung, D. Y., & Sumathy, K., (2007). A review and recent developments in photocatalytic water-splitting using TiO_2 for hydrogen production. *Renewable and Sustainable Energy Reviews, 11*(3), 401–425.

37. Kapdan, I. K., & Kargi, F., (2006). Bio-hydrogen production from waste materials. *Enzyme and Microbial Technology, 38*(5), 569–582.

38. Hallenbeck, P. C., & Benemann, J. R., (2002). Biological hydrogen production: fundamentals and limiting processes. *International Journal of Hydrogen Energy, 27*(11/12), 1185–1193.

39. Hawkes, F. R., Dinsdale, R., Hawkes, D. L., & Hussy, I., (2002). Sustainable fermentative hydrogen production: Challenges for process optimization. *International Journal of Hydrogen Energy, 27*(11/12), 1339–1347.

40. Haryanto, A., Fernando, S., Murali, N., & Adhikari, S., (2005). Current status of hydrogen production techniques by steam reforming of ethanol: A review. *Energy and Fuels, 19*(5), 2098–2106.

41. Zehner, O., (2012). *Green Illusions: The Dirty Secrets of Clean Energy and the Future of Environmentalism.* U of Nebraska Press.

42. McDowall, W., & Eames, M., (2006). Forecasts, scenarios, visions, back casts and roadmaps to the hydrogen economy: A review of the hydrogen futures literature. *Energy Policy, 34*(11), 1236–1250.

43. Zacharia, R., (2015). Review of solid state hydrogen storage methods adopting different kinds of novel materials. *Journal of Nanomaterials, 2015,* 4.

44. Das, L. M., (1996). On-board hydrogen storage systems for automotive application. *International Journal of Hydrogen Energy, 21*(9), 789–800.

45. https://www.energy.gov/eere/fuelcells/doe-technical-targets-onboard-hydrogen-storage-light-duty-vehicles (accessed on 3 August 2020).

46. Guo, Z. X., Shang, C., & Aguey-Zinsou, K. F., (2008). Materials challenges for hydrogen storage. *Journal of the European Ceramic Society, 28*(7), 1467–1473.

47. Zhevago, N. K., (2016). Other methods for the physical storage of hydrogen. In: *Compendium of Hydrogen Energy* (pp. 189–218). Wood head Publishing.

48. Satyapal, S., Petrovic, J., Read, C., Thomas, G., & Ordaz, G., (2007). The US department of energy's national hydrogen storage project: Progress towards meeting hydrogen-powered vehicle requirements. *Catalysis Today, 120*(3/4), 246–256.

49. Sakintuna, B., Lamari-Darkrim, F., & Hirscher, M., (2007). Metal hydride materials for solid hydrogen storage: A review. *International Journal of Hydrogen Energy, 32*(9), 1121–1140.

50. Klell, M., (2010). Storage of hydrogen in the pure form. *Handbook of Hydrogen Storage,* 187–214.

51. Schneemann, A., White, J. L., Kang, S., Jeong, S., Wan, L. F., Cho, E. S., Heo, T. W., et al., (2018). Nanostructured metal hydrides for hydrogen storage. *Chemical Reviews, 118*(22), 10775–10839.

52. Li, Q., & Bjerrum, N. J., (2002). Aluminum as anode for energy storage and conversion: A review. *Journal of Power Sources, 110*(1), 1–10.

53. Graetz, J., Reilly, J. J., Yartys, V. A., Maehlen, J. P., Bulychev, B. M., Antonov, V. E., Tarasov, B. P., & Gabis, I. E., (2011). Aluminum hydride as a hydrogen and energy storage material: Past, present, and future. *Journal of Alloys and Compounds, 509,* S517–528.

54. Pumera, M., & Wong, C. H., (2013). Graphane and hydrogenated graphene. *Chemical Society Reviews, 42*(14), 5987–5995.

55. Sevilla, M., & Mokaya, R., (2014). Energy storage applications of activated carbons: Super capacitors and hydrogen storage. *Energy and Environmental Science, 7*(4), 1250–1280.

56. Blankenship, II. T. S., Balahmar, N., & Mokaya, R., (2017). Oxygen-rich microporous carbons with exceptional hydrogen storage capacity. *Nature Communications, 8*(1), 1545.

57. Florusse, L. J., Peters, C. J., Schoonman, J., Hester, K. C., Koh, C. A., Dec, S. F., Marsh, K. N., & Sloan, E. D., (2004). Stable low-pressure hydrogen clusters stored in a binary clathrate hydrate. *Science, 306*(5695), 469–471.

58. Dillon, A., Jones, K. M., Bekkedahl, T. A., Kiang, C. H., Bethune, D. S., & Heben, M. J., (1997). Storage of hydrogen in single-walled carbon nanotubes. *Nature, 386*(6623), 377.

59. Zhou, H. C., Long, J. R., & Yaghi, O. M., (2012). *Introduction to Metal-Organic Frameworks.*

60. James, S. L., (2003). Metal-organic frameworks. *Chemical Society Reviews, 32*(5), 276–288.

61. Zhevago, N. K., Denisov, E. I., & Glebov, V. I., (2010). Experimental investigation of hydrogen storage in capillary arrays. *International Journal of Hydrogen Energy, 35*(1), 169–175.

62. Wicks, G. G., Heung, L. K., & Schumacher, R. F., (2008). SRNL's porous, hollow glass balls open new opportunities for hydrogen storage, drug delivery, and national defense. *American Ceramic Society Bulletin, 87*(6), 23–27.

63. Yu, X., Tang, Z., Sun, D., Ouyang, L., & Zhu, M., (2017). Recent advances and remaining challenges of nanostructured materials for hydrogen storage applications. *Progress in Materials Science, 88,* 1–48.

64. Hussein, A. K., (2015). Applications of nanotechnology in renewable energies: A comprehensive overview and understanding. *Renewable and Sustainable Energy Reviews, 42,* 460–476.

65. Roston, E., (2010). *The Carbon Age: How Life's Core Element Has Become Civilization's Greatest Threat.* Bloomsbury Publishing USA.

66. Zaytseva, O., & Neumann, G., (2016). Carbon nanomaterials: Production, impact on plant development, agricultural, and environmental applications. *Chemical and Biological Technologies in Agriculture, 3*(1), 17.

67. Kroto, H. W., Heath, J. R., O'Brien, S. C., Curl, R. F., & Smalley, R. E., (1985). *Nature, 318*(162), 1216.

68. Yadav, B. C., & Kumar, R., (2008). Structure, properties, and applications of fullerenes. *International Journal of Nanotechnology and Applications, 2*(1), 15–24.

69. Zhang, R., Zhang, Y., Zhang, Q., Xie, H., Qian, W., & Wei, F., (2013). Growth of half-meter long carbon nanotubes based on Schulz-Flory distribution. *ACS Nano, 7*(7), 6156–6161.

70. Li, D., & Kaner, R. B., (2008). Graphene-based materials. *Science, 320*(5880), 1170, 1171.
71. Ferrari, A. C., Meyer, J. C., Scardaci, V., Casiraghi, C., Lazzeri, M., Mauri, F., Piscanec, S., et al., (2006). Raman spectrum of graphene and graphene layers. *Physical Review Letters, 97*(18), 187401.
72. Tozzini, V., & Pellegrini, V., (2013). Prospects for hydrogen storage in graphene. *Physical Chemistry Chemical Physics, 15*(1), 80–89.
73. Narita, N., Nagai, S., Suzuki, S., & Nakao, K., (1998). Optimized geometries and electronic structures of graphyne and its family. *Physical Review B., 58*(16), 11009.
74. Jiao, Y., Du, A., Hankel, M., Zhu, Z., Rudolph, V., & Smith, S. C., (2011). Graphdiyne: A versatile nanomaterial for electronics and hydrogen purification. *Chemical Communications, 47*(43), 11843–11845.
75. Shunhong, Z., Jian, Z., Qian, W., Xiaoshuang, C., Yoshiyuki, K., & Puru, J., (2015). Penta-graphene: A new carbon allotrope. 7(2).
76. Lueking, A. D., Pan, L., Narayanan, D. L., & Clifford, C. E., (2005). Effect of expanded graphite lattice in exfoliated graphite nanofibers on hydrogen storage. *The Journal of Physical Chemistry B., 109*(26), 12710–12717.
77. Jin, H., Lee, Y. S., & Hong, I., (2007). Hydrogen adsorption characteristics of activated carbon. *Catalysis Today, 120*(3/4), 399–406.
78. Thomas, K. M., (2007). Hydrogen adsorption and storage on porous materials. *Catalysis Today, 120*(3/4), 389–398.
79. Chambers, A., Park, C., Baker, R. T., & Rodriguez, N. M., (1998). Hydrogen storage in graphite nanofibers. *The Journal of Physical Chemistry B., 102*(22), 4253–4256.
80. Liu, C., Fan, Y. Y., Liu, M., Cong, H. T., Cheng, H. M., & Dresselhaus, M. S., (1999). Hydrogen storage in single-walled carbon nanotubes at room temperature. *Science, 286*(5442), 1127–1129.
81. Texier-Mandoki, N., Dentzer, J., Piquero, T., Saadallah, S., David, P., & Vix-Guterl, C., (2004). Hydrogen storage in activated carbon materials: Role of the nanoporous texture. *Carbon, 12*(42), 2744–2747.
82. Gundiah, G., Govindaraj, A., Rajalakshmi, N., Dhathathreyan, K. S., & Rao, C. N., (2003). Hydrogen storage in carbon nanotubes and related materials. In: *Advances in Chemistry: A Selection of CNR Rao's Publications (1994–2003)* (pp. 305–309).
83. Gogotsi, Y., Dash, R. K., Yushin, G., Yildirim, T., Laudisio, G., & Fischer, J. E., (2005). Tailoring of nanoscale porosity in carbide-derived carbons for hydrogen storage. *Journal of the American Chemical Society, 127*(46), 16006, 16007.
84. Pupysheva, O. V., Farajian, A. A., & Yakobson, B. I., (2007). Fullerene nanocage capacity for hydrogen storage. *Nano Letters, 8*(3), 767–774.
85. Yoon, M., Yang, S., Wang, E., & Zhang, Z., (2007). Charged fullerenes as high-capacity hydrogen storage media. *Nano letters (Print), 7*(9), 2578–2583.
86. Zhong, Z. Y., Xiong, Z. T., Sun, L. F., Luo, J. Z., Chen, P., Wu, X., Lin, J., & Tan, K. L., (2002). Nanosized nickel (or cobalt)/graphite composites for hydrogen storage. *The Journal of Physical Chemistry B., 106*(37), 9507–9513.
87. Kuc, A., Zhechkov, L., Patchkovskii, S., Seifert, G., & Heine, T., (2007). Hydrogen sieving and storage in fullerene intercalated graphite. *Nano Letters, 7*(1), 1–5.
88. Lueking, A. D., Pan, L., Narayanan, D. L., & Clifford, C. E., (2005). Effect of expanded graphite lattice in exfoliated graphite nanofibers on hydrogen storage. *The Journal of Physical Chemistry B., 109*(26), 12710–12717.

89. Orimo, S., Majer, G., Fukunaga, T., Züttel, A., Schlapbach, L., & Fujii, H., (1999). Hydrogen in the mechanically prepared nanostructured graphite. *Applied Physics Letters, 75*(20), 3093–3095.

90. Tozzini, V., & Pellegrini, V., (2011). Reversible hydrogen storage by controlled buckling of graphene layers. *The Journal of Physical Chemistry C., 115*(51), 25523–25528.

91. Lin, Y., Ding, F., & Yakobson, B. I., (2008). Hydrogen storage by spillover on graphene as a phase nucleation process. *Physical Review B., 78*(4), 041402.

92. Zieliński, M., Wojcieszak, R., Monteverdi, S., Mercy, M., & Bettahar, M. M., (2005). Hydrogen storage on nickel catalysts supported on amorphous activated carbon. *Catalysis Communications, 6*(12), 777–783.

93. Im, J. S., Park, S. J., Kim, T., & Lee, Y. S., (2009). Hydrogen storage evaluation based on investigations of the catalytic properties of metal/metal oxides in electro spun carbon fibers. *International Journal of Hydrogen Energy, 34*(8), 3382–3388.

94. Dillon, A. C., Gennet, T., Alleman, J. L., Jones, K. M., Parilla, P. A., & Heben, M. J., (1999). *DOE Hydrogen Program, FY Progress Report.*

95. Smith, M. R., Bittner, E. W., Shi, W., Johnson, J. K., & Bockrath, B. C., (2003). Chemical activation of single-walled carbon nanotubes for hydrogen adsorption. *The Journal of Physical Chemistry B., 107*(16), 3752–3760.

96. Hirscher, M., Becher, M., Haluska, M., Dettlaff-Weglikowska, U., Quintel, A., Duesberg, G. S., Choi, Y. M., et al., (2001). Hydrogen storage in sonicated carbon materials. *Applied Physics A., 72*(2), 129–132.

97. Ye, Y., Ahn, C. C., Witham, C., Fultz, B., Liu, J., Rinzler, A. G., Colbert, D., et al., (1999). Hydrogen adsorption and cohesive energy of single-walled carbon nanotubes. *Applied Physics Letters, 74*(16), 2307–2309.

98. Nishimiya, N., Ishigaki, K., Takikawa, H., Ikeda, M., Hibi, Y., Sakakibara, T., Matsumoto, A., & Tsutsumi, K., (2002). Hydrogen sorption by single-walled carbon nanotubes prepared by a torch arc method. *Journal of Alloys and Compounds, 339*(1/2), 275–282.

99. Wu, X. B., Chen, P., Lin, J., & Tan, K. L., (2000). Hydrogen uptake by carbon nanotubes. *International Journal of Hydrogen Energy, 25*(3), 261–265.

100. Chen, Y., Shaw, D. T., Bai, X. D., Wang, E. G., Lund, C., Lu, W. M., & Chung, D. D., (2001). Hydrogen storage in aligned carbon nanotubes. *Applied Physics Letters, 78*(15), 2128–2130.

101. Tibbetts, G. G., Meisner, G. P., & Olk, C. H., (2001). Hydrogen storage capacity of carbon nanotubes, filaments, and vapor-grown fibers. *Carbon, 39*(15), 2291–2301.

102. Zhu, H., Cao, A., Li, X., Xu, C., Mao, Z., Ruan, D., Liang, J., & Wu, D., (2001). Hydrogen adsorption in bundles of well-aligned carbon nanotubes at room temperature. *Applied Surface Science, 178*(1–4), 50–55.

103. Kowalczyk, P., Hołyst, R., Terrones, M., & Terrones, H., (2007). Hydrogen storage in nanoporous carbon materials: Myth and facts. *Physical Chemistry Chemical Physics, 9*(15), 1786–1792.

104. Gallego, N. C., He, L., Saha, D., Contescu, C. I., & Melnichenko, Y. B., (2011). Hydrogen confinement in carbon nanopores: Extreme densification at ambient temperature. *Journal of the American Chemical Society, 133*(35), 13794–13797.

105. Ströbel, R., Jörissen, L., Schliermann, T., Trapp, V., Schütz, W., Bohmhammel, K., Wolf, G., & Garche, J., (1999). Hydrogen adsorption on carbon materials. *Journal of Power Sources, 84*(2), 221–224.

106. Fan, Y. Y., Liao, B., Liu, M., Wei, Y. L., Lu, M. Q., & Cheng, H. M., (1999). Hydrogen uptake in vapor-grown carbon nanofibers. *Carbon, 10*(37), 1649–1652.

107. De la Casa-Lillo, M. A., Lamari-Darkrim, F., Cazorla-Amoros, D., & Linares-Solano, A., (2002). Hydrogen storage in activated carbons and activated carbon fibers. *The Journal of Physical Chemistry B., 106*(42), 10930–10934.

108. Saha, D., & Deng, S., (2009). Hydrogen adsorption on ordered mesoporous carbons doped with Pd, Pt, Ni, and Ru. *Langmuir, 25*(21), 12550–12560.

109. Wang, L., & Yang, R. T., (2008). Hydrogen storage properties of carbons doped with ruthenium, platinum, and nickel nanoparticles. *The Journal of Physical Chemistry C., 112*(32), 12486–12494.

110. Zubizarreta, L., Menéndez, J. A., Job, N., Marco-Lozar, J. P., Pirard, J. P., Pis, J. J., Linares-Solano, A., et al., (2010). Ni-doped carbon xerogels for H_2 storage. *Carbon, 48*(10), 2722–2733.

111. Zubizarreta, L., Menéndez, J. A., Pis, J. J., & Arenillas, A., (2009). Improving hydrogen storage in Ni-doped carbon nanospheres. *International Journal of Hydrogen Energy, 34*(7), 3070–3076.

112. Hu, Z. L., Aizawa, M., Wang, Z. M., Yoshizawa, N., & Hatori, H., (2010). Synthesis and characteristics of graphene oxide-derived carbon nanosheet-Pd nanosized particle composites. *Langmuir, 26*(9), 6681–6688.

113. Zubizarreta, L., Arenillas, A., & Pis, J. J., (2009). Carbon materials for H_2 storage. *International Journal of Hydrogen Energy, 34*(10), 4575–4581.

114. Mpourmpakis, G., Tylianakis, E., & Froudakis, G. E., (2007). Carbon nanoscrolls: A promising material for hydrogen storage. *Nano Letters, 7*(7), 1893–1897.

115. Atkinson, K., Roth, S., Hirscher, M., & Grünwald, W., (2001). Carbon nanostructures: An efficient hydrogen storage medium for fuel cells. *Fuel Cells Bulletin, 4*(38), 9–12.

116. Ströbel, R., Garche, J., Moseley, P. T., Jörissen, L., & Wolf, G., (2006). Hydrogen storage by carbon materials. *Journal of Power Sources, 159*(2), 781–801.

117. Hirscher, M., & Becher, M., (2003). Hydrogen storage in carbon nanotubes. *Journal of Nanoscience and Nanotechnology, 3*(1, 2), 3–17.

118. Cheng, H. M., Yang, Q. H., & Liu, C., (2001). Hydrogen storage in carbon nanotubes. *Carbon, 39*(10), 1447–1454.

119. Hirscher, M., Becher, M., Haluska, M., Von, Z. F., Chen, X., Dettlaff-Weglikowska, U., & Roth, S., (2003). Are carbon nanostructures an efficient hydrogen storage medium? *Journal of Alloys and Compounds, 356*, 433–437.

120. Rzepka, M., Lamp, P., & Casa-Lillo, M. A. D. L., (1998). Physisorption of hydrogen on microporous carbon and carbon nanotubes. *The Journal of Physical Chemistry B., 102*(52), 10894–10898.

121. Wang, Q., & Johnson, J. K., (1999). Computer simulations of hydrogen adsorption on graphite nanofibers. *The Journal of Physical Chemistry B., 103*(2), 277–281.

122. Kubas, G. J., (2001). Metal-dihydrogen and σ-bond coordination: The consummate extension of the Dewar-Chatt-Duncanson model for metal-olefin π bonding. *Journal of Organometallic Chemistr., 635*(1/2), 37–68.

123. Skipper, C. V., (2013). *The Kubas Interaction in Transition Metal Based Hydrogen Storage Materials*. (Doctoral dissertation, UCL (University College London)).

124. Chung, C., Ihm, J., & Lee, H., (2015). Recent progress on Kubas-type hydrogen-storage nanomaterials: From theories to experiments. *Journal of the Korean Physical Society, 66*(11), 1649–1655.

125. Hoang, T. K., & Antonelli, D. M., (2009). Exploiting the Kubas interaction in the design of hydrogen storage materials. *Advanced Materials, 21*(18), 1787–1800.

126. Wang, L., & Yang, R. T., (2008). New sorbents for hydrogen storage by hydrogen spillover-a review. *Energy Environmental Science, 1*(2), 268–279.

127. Lueking, A. D., & Yang, R. T., (2004). Hydrogen spillover to enhance hydrogen storage-study of the effect of carbon physicochemical properties. *Applied Catalysis A: General, 265*(2), 259–268.

128. Cheng, H., Chen, L., Cooper, A. C., Sha, X., & Pez, G. P., (2008). Hydrogen spillover in the context of hydrogen storage using solid-state materials. *Energy and Environmental Science, 1*(3), 338–354.

129. Yoon, M., Yang, S., Hicke, C., Wang, E., Geohegan, D., & Zhang, Z., (2008). Calcium as the superior coating metal in functionalization of carbon fullerenes for high-capacity hydrogen storage. *Physical Review Letters, 100*(20), 206806.

130. Yildirim, T., & Ciraci, S., (2005). Titanium-decorated carbon nanotubes as a potential high-capacity hydrogen storage medium. *Physical Review Letters, 94*(17), 175501.

131. Kaur, M., & Pal, K., (2016). An investigation for hydrogen storage capability of zirconia-reduced graphene oxide nanocomposite. *International Journal of Hydrogen Energy, 41*(47), 21861–21869.

132. Parambhath, V. B., Nagar, R., & Ramaprabhu, S., (2012). Effect of nitrogen doping on hydrogen storage capacity of palladium decorated graphene. *Langmuir, 28*(20), 7826–7833.

133. Tylianakis, E., Psofogiannakis, G. M., & Froudakis, G. E., (2010). Li-doped pillared graphene oxide: A graphene-based nanostructured material for hydrogen storage. *The Journal of Physical Chemistry Letters, 1*(16), 2459–2464.

134. Lee, H., Ihm, J., Cohen, M. L., & Louie, S. G., (2010). Calcium-decorated graphene-based nanostructures for hydrogen storage. *Nano Letters, 10*(3), 793–798.

135. Ariharan, A., Viswanathan, B., & Nandhakumar, V., (2016). Heteroatom doped multi-layered graphene material for hydrogen storage application. *Graphene, 5*(02), 39.

136. Divya, P., & Ramaprabhu, S., (2014). Hydrogen storage in platinum decorated hydrogen exfoliated graphene sheets by spillover mechanism. *Physical Chemistry Chemical Physics, 16*(48), 26725–26729.

137. Samantaray, S. S., Sangeetha, V., Abinaya, S., & Ramaprabhu, S., (2018). Enhanced hydrogen storage performance in Pd3Co decorated nitrogen/boron doped graphene composites. *International Journal of Hydrogen Energy, 43*(16), 8018–8025.

138. Deng, W. Q., Xu, X., & Goddard, W. A., (2004). New alkali doped pillared carbon materials designed to achieve practical reversible hydrogen storage for transportation. *Physical Review Letters, 92*(16), 166103.

139. Ataca, C., Aktürk, E., Ciraci, S., & Ustunel, H., (2008). High-capacity hydrogen storage by metalized graphene. *Applied Physics Letters, 93*(4), 043123.

140. Ao, Z. M., & Peeters, F. M., (2010). High-capacity hydrogen storage in Al-adsorbed graphene. *Physical Review B., 81*(20), 205406.

141. Chen, P., Wu, X., Lin, J., & Tan, K. L., (1999). High H_2 uptake by alkali-doped carbon nanotubes under ambient pressure and moderate temperatures. *Science, 285*(5424), 91–93.

142. Yang, R. T., (2000). Hydrogen storage by alkali-doped carbon nanotubes-revisited. *Carbon, 38*(4), 623–626.

143. Pinkerton, F. E., Wicke, B. G., Olk, C. H., Tibbetts, G. G., Meisner, G. P., Meyer, M. S., & Herbst, J. F., (2000). Thermo gravimetric measurement of hydrogen absorption

in alkali-modified carbon materials. *The Journal of Physical Chemistry B., 104*(40), 9460–9467.

144. Prakash, J., Tripathi, B. M., Dasgupta, K., Chakravartty, J. K., Pai, M. R., Kumar, A., & Bharadwaj, S. R., (2016). Tuning of hydrogen storage property of Multi-walled carbon nanotube by decorating Ni, Cu, and Fe nanoparticles. *Current Nanomaterials, 1*(2), 124–131.

145. Vinayan, B. P., Sethupathi, K., & Ramaprabhu, S., (2012). Hydrogen storage studies of palladium decorated nitrogen doped graphene nanoplatelets. *Journal of Nanoscience and Nanotechnology, 12*(8), 6608–6614.

146. Reyhani, A., Mortazavi, S. Z., Mirershadi, S., Moshfegh, A. Z., Parvin, P., & Golikand, A. N., (2011). Hydrogen storage in decorated multi walled carbon nanotubes by Ca, Co, Fe, Ni, and Pd nanoparticles under ambient conditions. *The Journal of Physical Chemistry C., 115*(14), 6994–7001.

147. Sankaran, M., & Viswanathan, B., (2007). Hydrogen storage in boron substituted carbon nanotubes. *Carbon, 45*(8), 1628–1635.

148. Chen, C. H., & Huang, C. C., (2007). Hydrogen storage by KOH-modified multi-walled carbon nanotubes. *International Journal of Hydrogen Energy, 32*(2), 237–246.

149. Yadav, A., Faisal, M., Subramaniam, A., & Verma, N., (2017). Nickel nanoparticle-doped and steam-modified multiscale structure of carbon micro-nanofibers for hydrogen storage: Effects of metal, surface texture and operating conditions. *International Journal of Hydrogen Energy, 42*(9), 6104–6117.

150. Ariharan, A., Viswanathan, B., & Nandhakumar, V., (2016). *Phosphorous-Doped Porous Carbon Derived from Paste of Newly Growing Ficus benghalensis as Hydrogen Storage Material.*

151. Hwang, H. J., Kwon, Y., & Lee, H., (2012). Thermodynamically stable calcium-decorated graphyne as a hydrogen storage medium. *The Journal of Physical Chemistry C., 116*(38)20220–20224.

152. Nair, A. A., Sundara, R., & Anitha, N., (2015). Hydrogen storage performance of palladium nanoparticles decorated graphitic carbon nitride. *International Journal of Hydrogen Energy, 40*(8), 3259–3267.

153. Wang, Z., & Yang, R. T., (2010). Enhanced hydrogen storage on Pt-doped carbon by plasma reduction. *The Journal of Physical Chemistry C., 114*(13), 5956–5963.

154. Sankaran, M., & Viswanathan, B., (2006). The role of heteroatoms in carbon nanotubes for hydrogen storage. *Carbon, 44*(13), 2816–2821.

155. Yadav, A., Chakraborty, B., Gangan, A., Patel, N., Press, M. R., & Ramaniah, L. M., (2017). Magnetic moment controlling desorption temperature in hydrogen storage: A case of zirconium-doped graphene as a high capacity hydrogen storage medium. *The Journal of Physical Chemistry C., 121*(31), 16721–16730.

156. Li, C., Li, J., Wu, F., Li, S. S., Xia, J. B., & Wang, L. W., (2011). High capacity hydrogen storage in Ca decorated graphyne: A first-principles study. *The Journal of Physical Chemistry C., 115*(46), 23221–23225.

157. Chakraborty, B., Modak, P., & Banerjee, S., (2012). Hydrogen storage in yttrium-decorated single walled carbon nanotube. *The Journal of Physical Chemistry C., 116*(42), 22502–22508.

158. Yadav, M., & Xu, Q., (2012). Liquid-phase chemical hydrogen storage materials. *Energy and Environmental Science, 5*(12), 9698–9725.

159. Orimo, S. I., Nakamori, Y., Eliseo, J. R., Züttel, A., & Jensen, C. M., (2007). Complex hydrides for hydrogen storage. *Chemical Reviews, 107*(10), 4111–4132.
160. Thomas, K. M., (2007). Hydrogen adsorption and storage on porous materials. *Catalysis Today, 120*(3/4), 389–398.
161. Brandrup, J., Immergut, E. H., Grulke, E. A., Abe, A., & Bloch, D. R., (1999). *Polymer Handbook.* New York: Wiley.
162. Odian, G., (2004). *Principles of Polymerization.* John Wiley & Sons.
163. Billmeyer, F. W., & Billmeyer, F. W., (1984). *Textbook of Polymer Science.*
164. Wu, D., Xu, F., Sun, B., Fu, R., He, H., & Matyjaszewski, K., (2012). Design and preparation of porous polymers. *Chemical Reviews, 112*(7), 3959–4015.
165. Huang, J., & Turner, S. R., (2018). Hyper cross linked polymers: A review. *Polymer Reviews, 58*(1), 1–41.
166. McKeown, N. B., & Budd, P. M., (2006). Polymers of intrinsic microporosity (PIMs): Organic materials for membrane separations, heterogeneous catalysis, and hydrogen storage. *Chemical Society Reviews, 35*(8), 675–683.
167. Kato, R., & Nishide, H., (2018). Polymers for carrying and storing hydrogen. *Polymer Journal, 50*(1), 77.
168. Kato, R., Oya, T., Shimazaki, Y., Oyaizu, K., & Nishide, H., (2017). A hydrogen-storing quinaldine polymer: Nickel-electro deposition-assisted hydrogenation and subsequent hydrogen evolution. *Polymer International, 66*(5), 647–652.
169. Yamaguchi, R., Ikeda, C., Takahashi, Y., & Fujita, K. I., (2009). Homogeneous catalytic system for reversible dehydrogenation-hydrogenation reactions of nitrogen heterocycles with reversible interconversion of catalytic species. *Journal of the American Chemical Society, 131*(24), 8410–8412.
170. Kato, R., Yoshimasa, K., Egashira, T., Oya, T., Oyaizu, K., & Nishide, H., (2016). A ketone/alcohol polymer for cycle of electrolytic hydrogen-fixing with water and releasing under mild conditions. *Nature Communications, 7,* 13032.
171. Oyaizu, K., & Nishide, H., (2009). Radical polymers for organic electronic devices: A radical departure from conjugated polymers? *Advanced Materials, 21*(22), 2339–2344.
172. Kawahara, R., Fujita, K. I., & Yamaguchi, R., (2012). Dehydrogenative oxidation of alcohols in aqueous media using water-soluble and reusable Cp* Ir catalysts bearing a functional bipyridine ligand. *Journal of the American Chemical Society, 134*(8), 3643–3646.
173. Staudinger, H., & Wiedersheim, V., (1929). On high polymer compounds, Part 21:About the reduction of polystyrene. *Reports of the German Chemical Society (A and B Series), 62*(8), 2406–2411.
174. Mülhaupt, R., (2004). Hermann Staudinger and the origin of macromolecular chemistry. *Angewandte Chemie. International Edition, 43*(9), 1054–1063.
175. Germain, J., Hradil, J., Fréchet, J. M., & Svec, F., (2006). High surface area nanoporous polymers for reversible hydrogen storage. *Chemistry of Materials, 18*(18), 4430–4435.
176. Wood, C. D., Tan, B., Trewin, A., Niu, H., Bradshaw, D., Rosseinsky, M. J., Khimyak, Y. Z., et al., (2007). Hydrogen storage in micro porous hyper cross linked organic polymer networks. *Chemistry of Materials, 19*(8), 2034–2048.
177. McKeown, N. B., Gahnem, B., Msayib, K. J., Budd, P. M., Tattershall, C. E., Mahmood, K., Tan, S., et al., (2006). Towards polymer-based hydrogen storage materials: Engineering ultra micro porous cavities within polymers of intrinsic microporosity. *Angewandte Chemie International Edition, 45*(11), 1804–1807.

178. Budd, P. M., Butler, A., Selbie, J., Mahmood, K., McKeown, N. B., Ghanem, B., Msayib, K., et al., (2007). The potential of organic polymer-based hydrogen storage materials. *Physical Chemistry Chemical Physics, 9*(15), 1802–1808.

179. Ghanem, B. S., Msayib, K. J., McKeown, N. B., Harris, K. D., Pan, Z., Budd, P. M., Butler, A., et al., (2007). A triptycene-based polymer of intrinsic microposity that displays enhanced surface area and hydrogen adsorption. *Chemical Communications, 1,* 67–69.

180. Lee, J. Y., Wood, C. D., Bradshaw, D., Rosseinsky, M. J., & Cooper, A. I., (2006). Hydrogen adsorption in microporous hyper-crosslinked polymers. *Chemical Communications, 25,* 2670–2672.

181. Yuan, S., Kirklin, S., Dorney, B., Liu, D. J., & Yu, L., (2009). Nanoporous polymers containing stereo contorted cores for hydrogen storage. *Macromolecules, 42*(5), 1554–1559.

182. Šmigol, V., Švec, F., (1993). Preparation and properties of uniform beads based on macro porous glycidyl methacrylate-ethylene dimethacrylate copolymer: Use of chain transfer agent for control of pore-size distribution. *Journal of Applied Polymer Science, 48*(11), 2033–2039.

183. Fontanals, N., Marcé, R. M., Galià, M., & Borrull, F., (2003). Preparation and characterization of highly polar polymeric sorbents from styrene-divinylbenzene and vinylpyridine-divinylbenzene for the solid-phase extraction of polar organic pollutants. *Journal of Polymer Science Part A: Polymer Chemistry, 41*(13), 1927–1933.

184. Shaffei, K. A., Atta, A. M., Gomes, C. S., Gomes, P. T., El-Ghazawy, R. A., & Mahmoud, A. G., (2014). Preparation of polymer networks for hydrogen storage using the ullmann synthetic protocol. *Journal of Polymer Research, 21*(5), 445.

185. Germain, J., Svec, F., & Fréchet, J. M., (2008). Preparation of size-selective nanoporous polymer networks of aromatic rings: Potential adsorbents for hydrogen storage. *Chemistry of Materials, 20*(22), 7069–7076.

186. Thomas, A., Kuhn, P., Weber, J., Titirici, M. M., & Antonietti, M., (2009). Porous polymers: Enabling solutions for energy applications. *Macromolecular Rapid Communications, 30*(4/5), 221–236.

187. McKeown, N. B., Budd, P. M., & Book, D., (2007). Micro porous polymers as potential hydrogen storage materials. *Macromolecular Rapid Communications, 28*(9), 995–1002.

188. Ramimoghadam, D., Gray, E. M., & Webb, C. J., (2016). Review of polymers of intrinsic micro porosity for hydrogen storage applications. *International Journal of Hydrogen Energy, 41*(38), 16944–16965.

189. Yuan, S., Kirklin, S., Dorney, B., Liu, D. J., & Yu, L., (2009). Nanoporous polymers containing stereocorcorted cores of hydrogen storage. *Macromolecules, 42*(5), 1554–1559.

190. Rzepka, M., Lamp, P., & De La Casa-Lillo, M. A., (1998). Physisorption of hydrogen on microporous carbon and carbon nanotubes. *The Journal of Physical Chemistry B., 102*(52), 10894–10898.

191. Xu, Y., Jin, S., Xu, H., Nagai, A., & Jiang, D., (2013). Conjugated micro porous polymers: Design, synthesis, and application. *Chemical Society Reviews, 42*(20), 8012–8031.

192. Cooper, A. I., (2009). Conjugated micro porous polymers. *Advanced Materials, 21*(12), 1291–1295.

193. Dawson, R., Laybourn, A., Clowes, R., Khimyak, Y. Z., Adams, D. J., & Cooper, A. I., (2009). Functionalized conjugated micro porous polymers. *Macromolecules, 42*(22), 8809–8816.

194. Jiang, J. X., Wang, C., Laybourn, A., Hasell, T., Clowes, R., Khimyak, Y. Z., Xiao, J., et al., (2011). Metal-organic conjugated micro porous polymers. *Angewandte Chemie International Edition, 50*(5), 1072–1075.

195. Ding, S. Y., & Wang, W., (2013). Covalent organic frameworks (COFs), from design to applications. *Chemical Society Reviews, 42*(2), 548–568.

196. Feng, X., Ding, X., & Jiang, D., (2012). Covalent organic frameworks. *Chemical Society Reviews, 41*(18), 6010–6122.

197. Cote, A. P., Benin, A. I., Ockwig, N. W., O'Keeffe, M., Matzger, A. J., & Yaghi, O. M., (2005). Porous, crystalline, covalent organic frameworks. *Science, 310*(5751), 1166–1170.

198. El-Kaderi, H. M., Hunt, J. R., Mendoza-Cortés, J. L., Côté, A. P., Taylor, R. E., O'Keeffe, M., & Yaghi, O. M., (2007). Designed synthesis of 3D covalent organic frameworks. *Science, 316*(5822), 268–272.

199. Waller, P. J., Gándara, F., & Yaghi, O. M., (2015). Chemistry of covalent organic frameworks. *Accounts of Chemical Research, 48*(12), 3053–3063.

200. Kuhn, P., Antonietti, M., & Thomas, A., (2008). Porous, covalent triazine-based frameworks prepared by ionothermal synthesis. *Angewandte Chemie International Edition, 47*(18), 3450–3453.

201. Hug, S., Mesch, M. B., Oh, H., Popp, N., Hirscher, M., Senker, J., & Lotsch, B. V., (2014). A fluorene based covalent triazine framework with high CO_2 and H_2 capture and storage capacities. *Journal of Materials Chemistry A.,2*(16), 5928–5936.

202. Chen, X., Yuan, F., Gu, Q., & Yu, X., (2013). Light metals decorated covalent triazine-based frameworks as a high capacity hydrogen storage medium. *Journal of Materials Chemistry A., 1*(38), 11705–11710.

203. Ben, T., & Qiu, S., (2013). Porous aromatic frameworks: Synthesis, structure, and functions. *Cryst. Eng. Comm., 15*(1), 17–26.

204. Ben, T., Pei, C., Zhang, D., Xu, J., Deng, F., Jing, X., & Qiu, S., (2011). Gas storage in porous aromatic frameworks (PAFs). *Energy and Environmental Science, 4*(10), 3991–3999.

205. Lan, J., Cao, D., Wang, W., Ben, T., & Zhu, G., (2010). High-capacity hydrogen storage in porous aromatic frameworks with diamond-like structure. *The Journal of Physical Chemistry Letters, 1*(6), 978–981.

206. Lee, H., Choi, W. I., & Ihm, J., (2006). Combinatorial search for optimal hydrogen-storage nanomaterials based on polymers. *Physical Review Letters, 97*(5), 056104.

207. Durgun, E., Ciraci, S., Zhou, W., & Yildirim, T., (2006). Transition-metal-ethylene complexes as high-capacity hydrogen-storage media. *Physical Review Letters, 97*(22), 226102.

208. Li, A., Lu, R. F., Wang, Y., Wang, X., Han, K. L., & Deng, W. Q., (2010). Lithium-doped conjugated micro porous polymers for reversible hydrogen storage. *Angewandte Chemie International Edition, 49*(19), 3330–3333.

209. Sathe, R. Y., & Kumar, T. D., (2018). Paracyclophane functionalized with Sc and Li for hydrogen storage. *Chemical Physics Letters, 692*, 253–257.

210. Yuan, S., White, D., Mason, A., & Liu, D. J., (2013). Porous organic polymers containing carborane for hydrogen storage. *International Journal of Energy Research, 37*(7), 732–740.

211. Zhao, Y., Liu, F., Tan, J., Li, P., Wang, Z., Zhu, K., Mai, X., et al., (2019). Preparation and hydrogen storage of Pd/MIL-101 nanocomposites. *Journal of Alloys and Compounds, 772*, 186–192.

212. Kubas, G. J., (2009). Hydrogen activation on organometallic complexes and H$_2$ production, utilization, and storage for future energy. *Journal of Organometallic Chemistry, 694*(17), 2648–1653.

213. Ceperley, D. M., & Alder, B. J., (1980). Ground state of the electron gas by a stochastic method. *Physical Review Letters, 45*(7), 566.

214. Perdew, J. P., Burke, K., & Ernzerhof, M., (1996). Generalized gradient approximation made simple. *Physical Review Letters, 77*(18), 3865.

215. Martin, R. M., & Martin, R. M., (2004). *Electronic Structure: Basic Theory and Practical Methods.* Cambridge University Press.

216. Capelle, K., (2006). A bird's-eye view of density-functional theory. *Brazilian Journal of Physics, 36*(4A), 1318–1343.

217. Argaman, N., & Makov, G., (2000). Density functional theory: An introduction. *American Journal of Physics, 68*(1), 69–79.

218. Mayer, I., (2013). *Simple Theorems, Proofs, and Derivations in Quantum Chemistry.* Springer Science & Business Media.

219. Thijssen, J., (2007). *Computational Physics.* Cambridge university press.

220. Foulkes, W. M., Mitas, L., Needs, R. J., & Rajagopal, G., (2001). Quantum Monte Carlo simulations of solids. *Reviews of Modern Physics, 73*(1), 33.

221. Hammond, B. L., Lester, W. A., & Reynolds, P. J., (1994*). Monte Carlo Methods in Ab Initio Quantum Chemistry.* World Scientific.

222. Kolorenč, J., & Mitas, L., (2011). Applications of quantum Monte Carlo methods in condensed systems. *Reports on Progress in Physics, 74*(2), 026502.

223. Shavitt, I., & Bartlett, R. J., (2009). *Many-Body Methods in Chemistry and Physics: MBPT and Coupled-Cluster Theory.* Cambridge university press.

224. Jeziorski, B., & Monkhorst, H. J., (1981). Coupled-cluster method for multi determinantal reference states. *Physical Review A., 24*(4), 1668.

225. Sherrill, C. D., & Schaefer, III. H. F., (1999). The configuration interaction method: Advances in highly correlated approaches. In: *Advances in Quantum Chemistry* (Vol. 34, pp. 143–269). Academic Press.

226. Born, M., & Oppenheimer, R., (1927). To the quantum theory of molecules. *Annals of Physics, 389*(20), 457–484.

227. Hohenberg, P., & Kohn, W., (1964). Inhomogeneous electron gas. *Physical Review, 136*(3B), B864.

228. Kohn, W., & Sham, L. J., (1965). Self-consistent equations including exchange and correlation effects. *Physical Review, 140*(4A), A1133.

229. Becke, A. D., (1988). Density-functional exchange-energy approximation with correct asymptotic behavior. *Physical Review A., 38*(6), 3098.

230. Lee, C., Yang, W., & Parr, R. G., (1988). Development of the Collesalvetti correlation-energy formula into a functional of the electron density. *Physical Review B., 37*(2), 785.

231. https://www.msg.chem.iastate.edu/ (accessed on 3 August 2020).

232. http://gaussian.com/ (accessed on 3 August 2020).

233. https://www.vasp.at/ (accessed on 3 August 2020).

234. https://www.quantum-espresso.org/ (accessed on 3 August 2020).

235. http://susi.theochem.tuwien.ac.at/ (accessed on 3 August 2020).

236. http://elk.sourceforge.net/ (accessed on 3 August 2020).

237. Gangan, A., Chakraborty, B., Ramaniah, L. M., & Banerjee, S., (2019). First principles study on hydrogen storage in yttrium doped graphyne: Role of acetylene linkage in enhancing hydrogen storage. *International Journal of Hydrogen Energy*.

238. Henkelman, G., Arnaldsson, A., & Jónsson, H., (2006). A fast and robust algorithm for Bader decomposition of charge density. *Computational Materials Science, 36*(3), 354–360.

239. Mulliken, R. S., (1955). Electronic population analysis on LCAO-MO molecular wave functions. I. *The Journal of Chemical Physics, 23*(10), 1833–1840.

240. Kokalj, A., (1999). XCrySDen: A new program for displaying crystalline structures and electron densities. *Journal of Molecular Graphics and Modeling, 17*(3/4), 176–179.

241. Momma, K., & Izumi, F., (2011). VESTA 3 for three-dimensional visualization of crystal, volumetric, and morphology data. *Journal of Applied Crystallography, 44*(6), 1272–1276.

242. Grimme, S., (2006). Semi empirical GGA-type density functional constructed with a long-range dispersion correction. *Journal of Computational Chemistry, 27*(15), 1787–1799.

243. Dion, M., Rydberg, H., Schröder, E., Langreth, D. C., & Lundqvist, B. I., (2004). Van der Waals density functional for general geometries. *Physical Review Letters, 92*(24), 246401.

244. Lee, K., Murray, É. D., Kong, L., Lundqvist, B. I., & Langreth, D. C., (2010). Higher-accuracy van der Waals density functional. *Physical Review B., 82*(8), 081101.

245. Klimeš, J., Bowler, D. R., & Michaelides, A., (2009). Chemical accuracy for the van der Waals density functional. *Journal of Physics: Condensed Matter, 22*(2), 022201.

246. Klimeš, J., Bowler, D. R., & Michaelides, A., (2011). Van der Waals density functionals applied to solids. *Physical Review B., 83*(19), 195131.

247. Jiang, Y., Yang, S., Li, S., Liu, W., & Zhao, Y., (2015). Highly sensitive CO gas sensor from defective graphene: Role of van der Waals interactions. *Journal of Nanomaterials, 2015*, 5.

248. Chong, L., Li, J., Wu, F., Li, S., Xia, J., Wang, L., (2011). *J. Phys. Chem. C, 115*, 23221.

CHAPTER 10

Self-Assemblies of Macromolecular Systems Containing Green Polymers

ANDREEA IRINA BARZIC,[1] RALUCA MARINICA ALBU,[1] and CRISTIAN LOGIGAN[2]

[1]*"Petru Poni" Institute of Macromolecular Chemistry, Laboratory of Physical Chemistry of Polymers, 41A Grigore Ghica Voda Alley, Iasi – 700487, Romania*

[2]*Chemical Company SA, 14 Chemistry Bdv., Iasi – 700293, Romania*

ABSTRACT

Materials developed on the basis of green polymers are of major importance not only for the keeping a clean environment, but also for creating of advanced products with architecture controlled at micro- and nano-scale level. This chapter reviews the progress made in the field of self-assemblies in green macromolecular systems. The first short section presents some theoretical aspects and the most used synthesis and preparation routes of macromolecular assemblies of biopolymers. Then, the commonly employed characterization techniques of such soft nano-materials. The newest developments in green polymer self-assemblies are described emphasizing their practical importance. The presented research examples prove the range of possibilities and future trends in self-assembly processes that will significantly contribute to creation of materials having specific nano-architecture. The green materials, which is here under analysis, cover several application fields.

10.1 INTRODUCTION

Macromolecular self-assemblies are still a hot topic in materials and polymer science they facilitate the manufacturing of a variety of materials with extended applicability [1–7]. Preparation of soft nanomaterials in solution allows mimicking the exquisite structure and function of certain compounds found in nature, particularly biomacromolecules (proteins or DNA). Such nature inspired materials are outstanding from the point of view of their selectivity and precision, advancing applications in domains like information storage, transportation, and delivery. To achieve this purpose, it is paramount to utilize small molecules, amphiphiles, colloids, and polymers that enable fabrication of upgraded materials based on nanoscale blocks/units. To this end, multidisciplinary research must be conducted by combining skills of nanotechnology, polymer science, supramolecular chemistry, and biochemistry, but not limited to these ones.

In contrast to the macromolecules created and used by nature, synthetic polymers can be achieved from numerous monomers and combinations of them. Polymerization of these reactants enables different kinds of more or less complex homopolymers and copolymers. In this way, macromolecular engineering facilitates the access to uncommon architectures and shapes. Unfortunately, none of these macromolecular structures possess the sophistication and complexity noticed for those structures derived from the green materials (example: combination of 20 amino acids-natural monomers). The developments recorded in the past decades are mainly referring to refinements of procedures of polymer synthesis, allowing a higher control over the system's composition, molar mass, and entire architecture, however, they less address an assembly of structures of comparable complexity to proteins or other green polymers [2]. More recent reports [8, 9] provided another perspective by showing the possibility of diminishment tedious and time-consuming synthetic steps by engineering polymers of relatively small size that can self-organize via purely noncovalent forces (physical interactions). They are able to emulate the folding of peptide segments in proteins [10–14]. The most important aspect in such approaches is represented by the chemical structure of the synthetic macromolecules which must be able to transport all the information to direct the self-assembly process. Furthermore, self-assembly of polymer is a good and rapid pathway for the preparation of objects from nanometer to micrometer range, which are very hard if not impossible to be attained using the traditional chemical reactions. As a function of the resulted morphologies (size, shape, periodicity, etc.), these self-assembled materials are suitable for a huge number of applications, such as electronics [15], reusable elastomers [16], drug delivery [17], detergents [18], paints [19], and cosmetic products [20].

Having all these in view, the chapter present a short description of the theoretical aspects involved in macromolecular assemblies, followed by preparation and formation of such structures. The most employed characterization techniques of these soft materials are also described. The current approaches used in preparation of self-assemblies using green polymer are depicted in regard with the impact produced in various applicative fields.

10.2 THEORETICAL ASPECTS

The self-assembly of macromolecules in solution require some structural peculiarities. For instance this phenomenon is often encountered in block copolymers, which are containing covalently linked, and more recently non-covalently linked, macromolecular building blocks. The chemically distinct monomer units are combined in discrete blocks along the main chain. Depending on the used synthesis strategy and technique, one may obtain linear, branched, or cyclic molecular architectures. The compositional versatility of block-copolymers created innovative synthetic routes useful for producing previously unachievable levels of architectural complexity. In other words, self-assemblies of block copolymers are a good method for the attaining soft-matter-based core-shell nanoparticles (micelles) with extremely interesting properties and functions [2, 4, 5, 21–24].

A block copolymer with precise architecture is essential for controlling the solution self-assembly process by "adjusting" the interactions among the chain segments, both with each other, and the molecules of solvent [25]. Also, by tuning the molecular parameters (the degree of polymerization (N), the Flory-Huggins interaction parameter (χ), the, and the volume fraction (f)) it is possible to accomplish numerous types of morphologies, as experimentally and theoretically revealed [26, 27]. The architecture is also of paramount importance in controlling both the final morphologies and their extent of long-range order [28]. As a result of the exciting features of the block copolymers, the number of publications on this topic increased substantially since 1990s.

When placed in solvent media, block copolymers are suffering two main processes: micellization and gelation. The first one is taking place when the block copolymer is brought in contact with a large quantity of a selective solvent for one of the blocks. In such circumstances, the macromolecular chains are beginning to organize themselves in a variety of structures from micelles or vesicles to cylinders. The soluble block will be moving and orienting towards the continuous solvent medium and forms the 'corona' of the resulted micelle, while the insoluble part of the copolymer will be shielded from the solvent molecules in the 'core' of the structure. In contrast to micellization, gelation process appears in the semidilute to the high concentration domain of block copolymer solutions and results from the specific positioning of ordered micelles. The relative length of the blocks affects the micellar structure, namely when the soluble block prevails the micelles have a small core and a large corona (star micelles), while in the opposite situation the micelles have a short soluble corona (crew-cut micelles).

The analysis of a micellar system involves determining of several parameters, including the thermodynamic compatibility of the solvents, equilibrium constant, the critical micelle temperature (CMT), the critical micelle concentration (CMC), the molar mass M_w of the micelle, its aggregation number Z and the morphology. The mentioned variables influence the hydrodynamic radius (RH) R_H, the radius of gyration R_G, their ratio which impacts the micellar shape, the core radius R_C, and the thickness L of the corona. The shape and the dimensions of the aggregates are impacted by a numerous of parameters that influence the balance between three main forces acting on the system. They are related to the following aspects:

- The constraints between the blocks constituting the core, namely as a function of the type of solvent the block will be more or less stretched.
- The forces between chains forming the corona.
- The surface at the interface of the solvent/core of the micelle.

From a theoretical point of view, for the characterization of the aggregate structure is needed to account for the thermodynamic properties of self-assembly but also for the interactions among the macromolecules from the interior of aggregates. The last two factors (thermodynamics and intra-aggregate forces) together with the forces acting between different aggregates are influencing the type of self-assembled structure resulted in equilibrium conditions. In this context, it is necessary to comprehend the fundamentals that govern the relation between morphology and dimension of the aggregates acquired by self-assembly, including key factors such as concentration, composition, temperature, block length, copolymer architecture, and the nature of the solvents.

The theories developed to understand the behavior of block copolymers in solution can be categorized as follows:

- **Scaling Theory:** Proposed by de Gennes [29] estimated parameters such as the aggregation number or the radius for crew-cut micelles from specific information like the block length and interfacial tension. Another study [30] extended this theory for star micelles, while the report of Zhulina [31] based on this subject suggested that there are four types of micelles;
- **Self-Consistent Mean Field Theory:** Formulated by Noolandi and Hong [32] allow the evaluation of the dimension of spherical micelles at equilibrium, and the changes in the aggregation number as the degree of polymerization varies. Their theory is constructed by accounting for the molecular characteristics of the polymer, its concentration in solution, and prediction of the interfacial tension of the core/corona [33]. Further development of this approach enabled elucidation of other aspects, such as the evolution of the CMC with copolymer chemical structure (triblocks versus diblocks) [33], the temperature impact on the RH and aggregation number [34], the transition from spherical and cylindrical micellar systems [35], and the effects of polydispersity on CMC [36];
- **The Theory of Israelachvili and Coworkers [37]:** Employs geometrical considerations that foretell the micellization phenomenon and the corresponding formed morphologies. This model is applicable for both amphiphilic molecules of low molar and block copolymers. When discussing in the case of amphiphilic molecules one must consider the action of the forces governing the assembly into well defined structures: the hydrophobic attraction among insoluble hydrophobic segments, and the repulsion among the hydrophilic sequences due to electrostatic or steric interactions that both impose the contact with solvent of the amphiphilic molecules. When the attractive forces are prevailing, the interfacial area (a_o) per molecule will decrease, while if repulsive forces predominate, a_o will increase. The balance between such opposing forces is strongly affected by the geometry of the blocks, thus creating a variety of known morphologies. According to Israelachvili [37], the geometric properties are dictated by three parameters, namely the optimal interface a_o, the volume () taken by the hydrophobic segments, and the maximum length (l_c) of the chains. The described parameters are included in the definition of the packing parameter (p):

$$p = v/a_0 lc \qquad (10.1)$$

The shape factor p is < 1/3, 1, and > 1 for spherical micelles, bicontinuous bilayers, and inverted structures, respectively.

- **The Theory of Disher and Eisenberg [38]:** Attempted unify the experimental results reported in many investigations on amphiphilic block copolymers. They stated a rule for the appearance of polymersomes (polymer-based vesicles) in water: namely a ratio (denoted f) of the mass of the hydrophilic part to the total mass 35 ± 10%. The molecules having f > 45% will lead to micelles, while those having f < 25% will be able self-assemble into inverted structures.

10.3 PREPARATION OF MACROMOLECULAR SELF-ASSEMBLIES

The literature [2] indicates two main procedures for preparation of macromolecular self-assemblies in solution (micelles):

- Introduction of the copolymer in a good solvent for both blocks. In certain situations, the wanted micellar structure can be achieved by the manipulation of some solution parameters, such as temperature or a cosolvent. For other systems, the subsequent utilization of a selective solvent, into an already prepared solution in a common solvent it enables micelle formation.
- Removal of the common solvent from the system through a dialysis procedure. The approach to accomplish micelles is based on simply of inserting the dry solid copolymer powder into a selective solvent.

The choice for one or another method is mainly dictated by the system peculiarities. For instance, the glassy blocks like those from polystyrene (PS) generate the formation of highly stable micelles that are lacking an exchange between individual macromolecular chains and micelles ('frozen micelles'). For this situation, the first approach is more adequate.

10.4 METHODS OF CHARACTERIZATION OF MACROMOLECULAR SELF-ASSEMBLIES

10.4.1 ANALYSIS OF CMC

Fluorescence spectroscopy is one of the most common methods used in determination of the CMC [38, 39]. Pyrene is often preferred fluorescent probe given its strong fluorescence in nonpolar domains and its poor radiation in polar media. The shift of the excitation peak is monitored to prove the transfer of pyrene molecules into an increasingly nonpolar micellar medium. The ratio of intensities of the excitation maxima at 339 and 333 nm is changing with concentration. This leads to a plot with where a sudden variation of the data is noted. The crossover of the lines having distinct slopes represents the CMC value.

UV-absorption spectroscopy is another powerful technique for analysis of the CMC [40, 41]. The approach relies on the tautomerism of 1-phenyl-1,3-butadione between keto and enol forms which display distinct absorption maxima: 312 nm for the enolic form and 250 nm for the keto form, the former occurring in nonpolar solvents such as cyclohexane while the latter is formed in polar solvents, where H-bonding is disrupted in favor of the keto configuration [42].

There are other methods for CMC characterization which can be extended from small surfactants to block copolymer micelles, such as static or dynamic light is scattering or small-angle X-ray scattering (SAXS). However, there are some limitations of these techniques since signal intensity is very small considering the much lower CMC of block copolymers in regard with that of low molar mass surfactants.

10.4.2   DETERMINATION OF MICELLAR SIZE AND SHAPE

Specific micelle features (shape, size, and aggregation number Z) can be examined through several techniques. Depending on the chosen characterization method, a variety of data about the investigated system can be extracted. The most relevant are the following [2]:

- **Scattering Methods:**
  - **Static Light Scattering:** Technique to determine the average molar masses of self-assembled structures, including their CMC. Moreover, if scattering from the core and the corona of the micellar material is not so distinct one may estimate the RG value.
  - **Dynamic (or Quasi-Elastic) Light Scattering:** Allows analysis RH of a block copolymer micelles from the data regarding the diffusion coefficient. The sensitivity and versatility of this approach allow modifications in the micelle equilibrium owing to changes of temperature, pH, or other parameters.

- **Small-Angle X-Ray Scattering (SAXS):** Is adequate for the study of micellar solutions to attain overall and internal dimensions from dissimilarities in electron density of the solvent and the solute.
- **Small-Angle Neutron Scattering:** Provides significant data concerning the shape, and also the cross-section.
- **Non-Scattering Methods:**
- **Transmission Electron Microscopy:** Enable visualization of the micelle morphology, size or internal structure.
- **Atomic Force Microscopy:** Gives images from which the dimension, shape or mechanical resistance of the micelles can be determined.
- Dilute solution capillary.
- **Viscometry:** The changes in the solution time of flow are related to the micellization phenomenon.
- Membrane osmometry.
- Ultracentrifugation.
- Size exclusion chromatography.
- Nuclear magnetic resonance spectroscopy.

10.5   CURRENT DEVELOPMENTS IN SELF-ASSEMBLIES OF SYSTEMS CONTAINING GREEN POLYMERS

The association phenomenon in some systems containing a green polymer is briefly presented based on the latest advances in this research field. In this section of the chapter are reviewed the developments concerning cellulose, chitosan, and DNA.

10.5.1   CELLULOSE BASED SYSTEMS

The self-assembly of cellulose-based polyelectrolyte complexes (PECs) was investigated by Zhao et al. [43]. The procedure occurs in two stages:

- A weak anionic polyelectrolyte is dissolute in water and then is blended with four types of cationic polyelectrolytes (chitosan; cationic cellulose; poly-diallyl dimethylammonium chloride, PDDA; poly2-methacryloyloxy ethyl trimethylammonium chloride, PDMC) in HCl aqueous solutions; and

• The resulted PECs and then are re-dispersed in NaOH aqueous solutions.

Infrared spectroscopy data show that the ionic complexation indeed takes places between sodium carboxymethyl cellulose and the cationic polyelectrolytes and also probes the "two-step" method for obtaining PECs aqueous dispersions. PECs interfacial self-assembly process is interesting but not reported for other PECs dispersions processed in analogous conditions. Dynamic light scattering analyses reveal that R_H values of the PEC aggregates are comprised in the interval of 100–500 nm, while field emission electron scanning microscopy data indicate smaller values. Average R_H values of the PEC aggregates are ranging as follows: sodium carboxymethyl cellulose-cationic cellulose > sodium carboxymethyl cellulose-PDMC > sodium carboxymethyl cellulose-PDDA > sodium carboxymethyl cellulose-chitosan. From this, it can be extrapolated the idea that cationic polyelectrolytes with larger molecular weights produce larger PEC aggregates [44]. The basic constructing segments of PECs are ionically cross-linked and nano-sized PEC particles, which are amorphous considering their ionic cross-linking and "scrambled egg" structures [43]. The PECs dispersions in water and their component polyelectrolyte solutions are subjected to a drying process at 30°C on silicon supports. The corresponding FESEM data show tree-shaped and fractal surface patterns. Such morphologies are useful in surface patterning and membrane separation [43].

Cellulose nanocrystals (CNCs) are able to self-assembly [45, 46] having a remarkable ability to organize into photonic structures. The colloidal stability of these materials is a mandatory to achieve self-assembly, but in the same time, insufficiently charged CNCs (such as isolated after hydrolysis) also impede the possibility to self-assemble. After preparation of the suspensions are subjected to purification through dialysis and then introduced in polar solvents [47]. If a surfactant [48] or further functionalization [49, 50] is applied, colloidal stability is extended towards extreme conditions.

Hata et al. [51] studied the oligomerization-induced self-assembly of crystalline cellulose into nanoribbon networks supported by organic solvents. The controlled macromolecular self-assembly into organized hierarchical structures using a simple method is hard, but they showed that organic solvents can serve as small-molecule additives and thus enable the control of the self-assembly of cellulose oligomers into the desired network structures. They used cellodextrin phosphorylase-catalyzed oligomerization of phosphorylated glucose monomers from D-glucose primers, which generate precipitates of nanosheet-shaped crystals in water. The introduction of water-miscible organic solvents-dimethyl sulfoxide (DMSO), N,N-dimethylformamide (DMF), acetonitrile (MeCN), and ethanol (EtOH)-favor occurrence of nanoribbon networks and avoid the irregular aggregation and the precipitation of the nanosheets. The results prove that small-molecule additives are useful for controlling the self-assembly of crystalline oligosaccharides for the fabrication of hierarchically structured materials with large robustness in a facile manner.

10.5.2   CHITOSAN BASED SYSTEMS

The self-assembly of enzymatic hydrolysates of chitosan and carboxymethyl cellulose allows formation of biocompatible particles [52]. Both polymers were hydrolyzed with chitosanase and cellulase respectively to attain fragments of lower molecular weights. Polymer nanoparticles are almost instantly formed only by blending the two hydrolysate solutions. The particle size distribution resulted to be relatively narrow, around 200 nm in mean size. The mean particle dimension diminished from 226 nm to 165 nm as the molecular weight of chitosan hydrolysate was lowered from 9.5 to 6.8 kDa. The composition of the system also impacted the particle size. The blended polymer nanoparticles are found to be stably suspended over 1 week even for small pH (pH 3.0) conditions, high ionic strength (NaCl 1 M), or temperatures around 4°C.

Brunel et al. [53] obtained nanohydrogels of pure chitosan, which are lacking any toxic solvent or chemical cross-linker. This can be achieved by an ammonia-generated physical gelation of a reverse emulsion of a chitosan/triglyceride mixture. The produced colloids are accomplished with a controlled distribution of dimensions and present a positive surface charge. Such macromolecular assemblies opened new perspectives toward novel nano-carriers for bioactive compounds. Chondroitin sulfate renders PECs having positively charged surface and thus determining negative chitosan-based colloidal hydrogels that maintain of the original average size of the dispersion. The mode of assembly of HIV-1 p24 protein with the formed colloids is mainly based on multiple interactions among the nanohydrogels and the protein, regardless of their surface charges. In any case, the amounts of introduced protein are not so high, suggesting a surface association. For comparison, the assembly of an immunoglobulin (IgG) is studied and led to different data from p24. No association is noted with the positive colloidal hydrogels while a larger loading capacity could be attained with the negative ones. The reported fully biodegradable submicrometric physical hydrogels based on green polymers could cargo a several types of biomolecules recommending them as versatile carrier's high impact in biomedicine [54].

On the other hand, the chitosan-based nanoparticles, prepared via self-assembly, can be made to become sensitive to some stimuli [55]. Chitosan-graft-poly(N-isopropylacrylamide) copolymers (CS-g-PNIPAAm) were obtained by polymerization of NIPAAm monomer in the presence of chitosan in water, where cerium ammonium nitrate has the role of initiator. The copolymer solution was further diluted using deionized water and subjected to a higher temperature in order to ensure the occurrence of the self-assembly. Based on this procedure micelles of CS-g-PNIPAAm are attained. For reinforcement of the desired micelle structure and to convert them into nanoparticles, in the system is introduced glutaraldehyde. Transmission electron microscopy of samples showed a porous or hollow structure of nanoparticles. The dimensions of the nanoparticles can be tuned thought the temperature of the environment and can be applied in hydrophilic drug delivery.

Another report on sensitive macromolecular assemblies based on chitosan is made by Chen et al. [56]. They prepared an innovative pH-sensitive PEC for insulin designed for oral administration which result after self-assembly of two oppositely charged nanoparticles of 200–300 nm (chitosan covered nanoparticles and alginate covered nanoparticles) via electrostatic interaction and optimized double emulsion procedure. The morphology of the PEC system is pH-sensitive and the reasons for the morphological variation arise from the strength of the electrostatic interaction among the green polymer and alginate at several pH values. The PEC materials display non-cytotoxicity against Caco-2 cell. The insulin-loaded systems are effective in decreasing the insulin level in rats and keep prolonged release of the active substance.

Chitosan/cellulose systems are able to self-assemble into a 3D porous structure resulting fascinating materials [57]. Cellulose extracted from plants and chitosan taken from marine sources was used by Zhang and collaborators [57] together with a clear dissolution strategy. The renewable materials are dispersed into a homogeneous solution in green alkali/urea media. Given the hydrogen-bond interaction among the green polymers, they were able to self-assemble into a unique 3D alloy structure. The high amount of interface interactions enhanced the tensile strength and breaking elongation of the alloy sample up to 69%–260%, in regard to pure chitosan film. The 3D alloy materials present a pore diameter of 920–1435 nm together with an outstanding high cell viability that exceeds 95%. Moreover, accelerated wound healing rate of 95.6% is attained in the presence of the alloy sample after 14 days using a full-thickness wound in a rat.

10.5.3 DNA BASED SYSTEMS

DNA is a widely studied biopolymer which in certain conditions leads to outstanding self-assembled structures. The self-assembly of micellar or vesicular structures in DNA-polymer conjugates was reviewed by Kedracki and co-workers [58]. When a hydrophobic polymer sequence is coupled to a nucleotide unit it is possible self-assemble in solution rendering to core-shell micelles made of a hydrophobic core covered by a hydrophilic corona of the DNA fragment. DNA-polymer conjugates that form micelles for the targeting of antisense oligonucleotide to cells are studied [59]. Another category of amphiphilic DNA-polymer conjugates which generate micellar structures in water solution are suitable for molecular recognition to hybridize them with gold nanoparticles, leading to the construction of stimuli-responsive materials [60].

The self-assembly in DNA-based systems can be used not only in drug delivery but also in imaging applications. Peterson and Heemstra [61] prepared micelles and vesicles via phase-driven self-assembly of monomer structures containing discrete hydrophilic and hydrophobic sequences. In these materials, DNA plays the role of the hydrophilic block of the amphiphilic monomers and facilitates assemblies with the advantage of the good information storage and ability of molecular recognition.

By altering monomer structure, one may tune the morphology of DNA-polymer conjugates and monitor the influence of assembly on DNA stability and hybridization.

Another investigation developed by Alemdaroglu [62] the DNA is coupled to a poly(propylene oxide) (PPO) of higher biocompatibility than other macromolecular compounds used for the synthesis of DNA-polymer conjugates. In aqueous solution, the DNA-b-PPO copolymers begin to self-assemble resulting micelles composed of a hydrophobic PPO core and a shell made from nucleotide segments.

These materials are suitable as nanoreactors to carry out organic reactions [62]. Moreover, in other investigation it was proved that DNA-based micelles can suffer a morphological transition from spherical to rod-like structures after hybridization with long DNA segments, which were selected in such a way that they codify several times the complementary part of the DNA representing the self-assembling conjugate [63].

Aside from the self-organization of DNA-polymer systems into micelles, the occurrence of vesicular structures has been also reported [64, 65]. The vesicles are prepared by the self-assembly of an amphiphilic DNA-polymer conjugate leading to the coupling among a highly hydrophobic macromolecular segment and a DNA unit. The formation of vesicular structures is of great importance for some pharmaceutical and biomedical applications because they are able to be filled with substances which can be specifically delivered to cells. Vebert-Nardin and colleagues [64, 65] showed for the first time the self-organization of vesicular structures self-assembled from DNA-polymer system via reaction of a suitable polymer segment to a 12 nucleotide sequence.

In recent trends [66], the scientists are focused on self-assembled materials with upgraded programmable features made from DNA-based systems. There are several strategies of directing the self-organization of DNA-inspired, sequence-specific polyphosphodiesters into structures with unique characteristics.

This highlights the fact that DNA-derived systems are not fully exploited in solving some issues in certain technological areas, particularly controlled polymer folding, and assembly. The accurate positioning of structurally distinct molecules within a macromolecular chain renders unmatched possibilities for encoding information on the molecular level and broadcast it to the microscopic and even macroscopic scale through the noncovalent interactions.

10.6 CONCLUDING REMARKS

The chapter aimed at presenting an up-to-date report on the developments made in the area of self-assemblies in the systems containing green polymers. The theoretical fundaments that lie at the basis of organization of macromolecules in solutions are described. The most used methods employed in preparation and characterization of self-assembled structures is presented. The most significant researches involving green polymer self-organization and the practical importance are also depicted. The control of the shape and size of such nano-objects and the construction of particular properties are no longer unreachable challenges. This research domain offers opportunities for chemists, physicists, and biochemists, to create systems that might eventually mimic in sophistication and accuracy the biological structures developed by nature.

ACKNOWLEDGMENTS

The financial support of European Social Fund for Regional Development, Competitiveness Operational Programme Axis 1-Project "Petru Poni Institute of Macromolecular Chemistry-Interdisciplinary Pol for Smart Specialization through Research and Innovation and Technology Transfer in Bio(nano)polymeric Materials and (Eco)Technology," InoMatPol (ID P_36_570, Contract 142/10.10.2016, cod MySMIS: 107464) is gratefully acknowledged.

KEYWORDS

- **cellulose nanocrystals**
- **critical micelle concentration**
- **green polymers**
- **immunoglobulin**
- **organic solvents**
- **self-assemblies**

REFERENCES

1. Zhulina, E. B., & Borisov, O. V., (2002). Self-assembly in solution of block copolymers with annealing polyelectrolyte blocks. *Macromolecules, 35*, 9191–9203.
2. Rodrıguez-Hernandez, J., Checot, F., Gnanou, Y., & Lecommandoux, S., (2005). Toward 'smart' nano-objects by self-assembly of block copolymers in solution. *Prog. Polym. Sci., 30*, 691–724.
3. Patterson, J. P., Robin, M. P., Chassenieux, C., Colombani, O., & O'Reilly, R. K., (2014). The analysis of solution self-assembled polymeric nanomaterials. *Chem. Soc. Rev., 43*, 2412–2425.
4. Huang, F., O'Reilly, R., & Zimmerman, S. C., (2014). Polymer self-assembly: A web themed issue. *Chem. Commun., 50*, 13415–13416.
5. Tritschler, U., Pearce, S., Gwyther, J., Whittell, G. R., & Manners, I., (2017). 50th Anniversary perspective: Functional nanoparticles from the solution self-assembly of block copolymers. *Macromolecules, 50*, 3439–3463.
6. Feng, H., Lu, X., Wang, W., Kang, N. G., & Mays, J. W., (2017). Block copolymers: Synthesis, self-assembly, and applications. *Polymers, 9*, 494–520.
7. Ariga, K., Nishikawa, M., Mori, T., Takeya, J., Shrestha, L. K., & Hill, J. P., (2019). Self-assembly as a key player for materials nanoarchitectonics. *Sci. Technol. Adv. Mat., 20*, 51–95.

8. Van Rijn, P., (2013). Polymer directed protein assemblies. *Polymers, 5*, 576–599.

9. Luo, Q., Dong, Z., Hou, C., & Liu, J., (2014). Protein-based supramolecular polymers: Progress and prospect. *Chem. Commun., 50*, 9997–10007.

10. Hendricks, M. P., Sato, K., Palmer, L. C., & Stupp, S. I., (2017). Supramolecular assembly of peptide. *Acc. Chem. Res., 50*, 2440–2448.

11. Cavalli, S., Albericio, F., & Kros, A., (2010). Amphiphilic peptides and their cross-disciplinary role as building blocks for nanoscience. *Chem. Soc. Rev., 39*, 241–263.

12. Rho, J. Y., Cox, H., Mansfield, E. D. H., Ellacott, S. H., Peltier, R., Brendel, J. C., Hartlieb, M., et al., (2019). Dual self-assembly of supramolecular peptide nanotubes to provide stabilization in water. *Nat. Commun., 10*, 4708.

13. Raymond, D. M., & Nilsson, B. L., (2018). Multi component peptide assemblies. *Chem. Soc. Rev., 47*, 3659–3720.

14. Ottera, R., & Besenius, P., (2019). Supramolecular assembly of functional peptide-polymer conjugates. *Org. Biomol. Chem., 17*, 6719–6734.

15. Edrington, A. C., Urbas, A. M., DeRege, P., Chen, C., Swager, T. M., Hadjichristidis, M., et al., (2001). Polymer-based photonic crystals. *Adv. Mater., 13*, 421–425.

16. Holden, G., Legge, N. R., Quirck, R., & Schroeder, II. E., (1996). *Thermoplastic Elastomers* (2ndedn.). Cincinnati: Honser/Gardner Publication.

17. Wang, D., Tong, G., Dong, R., Zhou, Y., Shen, J., & Zhu, X., (2014). Self-assembly of supramolecularly engineered polymers and their biomedical applications. *Chem. Commun., 50*, 11994–12017.

18. Sundararajan, P. R., (2017). *Physical Aspects of Polymer Self-Assembly*. Wiley: USA.

19. Baradie, B., & Schoichet, M. S., (2002). Synthesis of fluorocarbon-vinyl acetate copolymers in supercritical carbon dioxide: Insight into bulk properties. *Macromolecules, 35*, 3569–3575.

20. Nakai, S., Nakai, A., & Michida, T., (2016). Microencapsulation of ascorbic acid for cosmetic by utilizing self-assembly of phase separated polymer. *Chem. Pharm. Bull., 64*, 1514–1518.

21. Rodríguez-Hernández, J., Chécot, F., Gnanou, Y., & Lecommandoux, S., (2005). Toward 'smart' nano-objects by self-assembly of block copolymers in solution. *Prog. Polym. Sci., 30*, 691–724.

22. Ge, Z., & Liu, S., (2013). Functional block copolymer assemblies responsive to tumor and intracellular microenvironments for site specific drug delivery and enhanced imaging performance. *Chem. Soc. Rev., 42*, 7289–7325.

23. Kempe, K., Wylie, R. A., Dimitriou, M. D., Tran, H., Hoogenboom, R., Schubert, U. S., Hawker, C. J., Campos, L. M., & Connal, L. A., (2016). Preparation of non-spherical particles from amphiphilic block copolymers. *J. Polym. Sci., Part A: Polym. Chem., 54*, 750–757.

24. Kataoka, K., Harada, A., & Nagasaki, Y., (2001). Block copolymer micelles for drug delivery: Design, characterization, and biological significance. *Adv. Drug Delivery Rev., 47*, 113–131.

25. Mai, Y., & Eisenberg, A., (2012). Self-assembly of block copolymers. *Chem. Soc. Rev., 41*, 5969–5985.

26. Lennon, E. M., Katsov, K., & Fredrickson, G. H., (2008). Free energy evaluation in field-theoretic polymer simulations. *Phys. Rev. Lett., 101*, 138302.

27. Matsen, M. W., & Bates, F. S., (1996). Origins of complex self-assembly in block copolymers. *Macromolecules, 29*, 7641–7644.

28. Poelma, J. E., Ono, K., Miyajima, D., Aida, T., Satoh, K., & Hawker, C. J., (2012). Cyclic block copolymers for controlling feature sizes in block copolymer lithography. *ACS Nano, 6*, 10845–10854.

29. De Gennes, P. G., (1978). In: Libert, L., (ed.), *Solid State Physics*. Academic Press: New York.

30. Daoud, M., & Cotton, J. P., (1982). Star-shaped polymers: A model for the conformation and its concentration dependence. *J. Phys., 43*, 531–538.

31. Zhulina, E. B., & Birshtein, T. M., (1985). Conformations of molecules of block copolymers in selective solvents (micellar structures). *Vysokomol. Soed., Seriya A.,27*, 511–517.

32. Noolandi, J., & Hong, K. M., (1983). Theory of block copolymer micelles in solution. *Macromolecules, 16*, 1443–1448.

33. Linse, P., (1993). Micellization of poly(ethylene oxide)-poly(propylene oxide) block copolymers in aqueous solution. *Macromolecules, 26*, 4437–4449.

34. Linse, P., & Malmsten, M., (1992). Temperature-dependent micellization in aqueous block copolymer solutions. *Macromolecules, 25*, 5434–5439.

35. Linse, P., (1994). Micellization of poly(ethylene oxide)-poly(propylene oxide) block copolymers in aqueous solution: Effect of polymer polydispersity. *Macromolecules, 27*, 6404–6417.

36. Gao, Z., & Eisenberg, E., (1993). A model of micellization for block copolymers in solutions. *Macromolecules, 26*, 7353–7360.

37. Israelachvili, J. N., (1992). *Intermolecular and Surface Forces*. London: Harcourt Brace and Company.

38. Lupu, M., Macocinschi, D., Epure, V., Ioanid, A., Grigoras, C. V., & Ioan, S., (2006). Micellization process in poly(ether urethane) solutions. *J. Macromol. Sci. Part B, 45*, 395–405.

39. Kalyanasundaram, K., & Thomas, J. K., (1977). Environmental effects on vibronic band intensities in pyrene monomer fluorescence and their application in studies of micellar systems. *J. Am. Chem. Soc., 99*, 2039–2044.

40. Geng, J., Johnson, B. F. G., Wheatley, A. E. H., & Luo, J. K., (2014). Spectroscopic route to monitoring individual surfactant ions and micelles in aqueous solution: A case study. *Cent. Eur. J. Chem., 12*, 307–311.

41. Tanhaei, B., Saghatoleslami, N., Chenar, M. P., Ayati, A., Hesampour, M., & Mänttäri, M., (2012). Experimental study of CMC evaluation in single and mixed surfactant systems, using the UV-Vis spectroscopic method. *J. Surfactants Deterg., 16*, 357–362.

42. Dominguez, A., Fernández, A., González, N., Iglesias, E., & Montenegro, L., (1997). Determination of critical micelle concentration of some surfactants by three techniques. *J. Chem. Education, 74*, 1227–1231.

43. Zhao, Q., Qian, J., Gui, Z., An, Q., & Zhu, M., (2010). Interfacial self-assembly of cellulose-based polyelectrolyte complexes: Pattern formation of fractal "trees." *Soft Matter, 6*, 1129–1137.

44. Schatz, C., Bionaz, A., Lucas, J. M., Pichot, C., Viton, C., Domard, A., & Delair, T., (2005). Formation of polyelectrolyte complex particles from self-complexation of N-sulfated chitosan. *Biomacromolecules, 6*, 1642–1647.

45. Parker, R. M., Guidetti, G., Williams, C. A., Zhao, T., Narkevicius, A., Vignolini, S., & Frka-Petesic, B., (2018). The self-assembly of cellulose nanocrystals: Hierarchical design of visual appearance. *Adv. Mater., 30*, 1704477.

46. Rofouie, P., Alizadehgiashi, M., Mundoor, H., Smalyukh, I. I., & Kumacheva, E., (2018). Self-assembly of cellulose nanocrystals into semi-spherical photonic cholesteric films. *Adv. Funct. Mater., 28,* 1803852.

47. Bruckner, J. R., Kuhnhold, A., Honorato-Rios, C., Schilling, T., & Lagerwall, J. P. F., (2016). Enhancing self-assembly in cellulose nanocrystal suspensions using high-permittivity solvents. *Langmuir, 32,* 9854–9862.

48. Salajková, M., Berglund, L. A., & Zhou, Q., (2012). Hydrophobic cellulose nanocrystals modified with quaternary ammonium salts. *J. Mater. Chem., 22,* 19798–19805.

49. Tang, J., Sisler, J., Grishkewich, N., & Tam, K. C., (2017). Functionalization of cellulose nanocrystals for advanced applications. *J. Colloid Interface Sci., 494,* 397–409.

50. Natterodt, J. C., Petri-Fink, A., Weder, C., & Zoppe, J. C., (2017). Cellulose nanocrystals: Surface modification, applications, and opportunities at interfaces. *Int. J. Chem., 71,* 376–383.

51. Hata, Y., Fukaya, Y., Sawada, T., Nishiura, M., & Serizawa, T., (2019). Biocatalytic oligomerization-induced self-assembly of crystalline cellulose oligomers into nanoribbon networks assisted by organic solvents. *Beilstein J. Nanotechnol., 10,* 1778–1788.

52. Ichikawa, Iwamoto, S., & Watanabe, J., (2005). Formation of biocompatible nanoparticles by self-assembly of enzymatic hydrolysates of chitosan and carboxymethyl cellulose. *Biosci. Biotechnol. Biochem., 69,* 1637–1642.

53. Brunel, F., Véron, L., David, L., Domard, A., Verrier, B., & Delair, T., (2010). Self-assemblies on chitosan nanohydrogels. *Macromol. Biosci., 10,* 424–432.

54. Quiñones, J. P., Peniche, H., & Peniche, C., (2018). Chitosan based self-assembled nanoparticles in drug delivery. *Polymers, 10,* 235.

55. Chuang, C. Y., Don, T. M., & Chiu, W. Y., (2011). Preparation of environmental-responsive chitosan-based nanoparticles by self-assembly method. *Carbohydr. Polym., 84,* 765–769.

56. Chen, T., Li, S., Zhu, W., Liang, Z., & Zeng, Q., (2019). Self-assembly pH-sensitive chitosan/alginate coated polyelectrolyte complexes for oral delivery of insulin. *J. Microencapsul., 36,* 96–107.

57. Zhang, R., Xie, J., Yang, B., Fu, F., Tang, H., Zhang, J., Zhao, Y., et al., (2019). Self-assembly of chitosan and cellulose chains into a 3D porous polysaccharide alloy films: Co-dissolving, structure, and biological properties. *Appl. Surf. Sci., 493,* 1032–1041.

58. Kedracki, D., Safir, I., Gour, N., Ngo, K. X., & Vebert-Nardin, C., (2013). DNA-polymer conjugates: From synthesis, through complex formation and self-assembly to applications. *Adv. Polym. Sci., 253,* 115–150.

59. Jeong, J. H., & Park, T. G., (2001). Novel polymer-DNA hybrid polymeric micelles composed of hydrophobic poly(D,L-lactic-co-glycolic acid) and hydrophilic oligonucleotides. *Bioconjug. Chem., 12,* 917–923.

60. Li, Z., Zhang, Y., Fullhart, P., & Mirkin, C. A., (2004). Reversible and chemically programmable micelle assembly with DNA block-copolymer amphiphiles. *Nano Lett., 4,* 1055–1058.

61. Peterson, A. M., & Heemstra, J. M., (2001). Controlling self-assembly of DNA-polymer conjugates for applications in imaging and drug delivery. *WIREs Nanomed. Nanobi., 7,* 282–297.

62. Alemdaroglu, F. E., Ding, K., Berger, R., & Herrmann, A., (2006). DNA-templated synthesis in three dimensions: Introducing a micellar scaffold for organic reactions. *Angew. Chem. Int. Ed. Engl., 45,* 4206–4210.

63. Ding, K., Alemdaroglu, F. E., Borsch, M., Berger, R., & Herrmann, A., (2007). Engineering the structural properties of DNA block copolymer micelles by molecular recognition. *Angew. Chem. Int. Ed. Engl., 46*, 1172–1175.

64. Teixeira, F. Jr., Rigler, P., & Vebert-Nardin, C., (2007). Nucleo-copolymers: Oligonucleotide-based amphiphilic dib lock copolymers. *Chem. Commun.*, 1130–1132.

65. Vebert-Nardin, C., (2011). Towards biologically active self-assemblies: Model nucleotide chimeras. *Chimia., 65*, 782–786.

66. Vybornyi, M., Vyborna, Y., & Haner, R., (2019). DNA-inspired oligomers: From oligophosphates to functional materials. *Chem. Soc. Rev., 48*, 4347–4360.

Index